SpringerBriefs in Physics

SpringerBriefs in Physics are a series of slim high-quality publications encompassing the entire spectrum of physics. Manuscripts for SpringerBriefs in Physics will be evaluated by Springer and by members of the Editorial Board. Proposals and other communication should be sent to your Publishing Editors at Springer.

Featuring compact volumes of 50 to 125 pages (approximately 20,000–45,000 words), Briefs are shorter than a conventional book but longer than a journal article. Thus, Briefs serve as timely, concise tools for students, researchers, and professionals.

Typical texts for publication might include:

- A snapshot review of the current state of a hot or emerging field
- A concise introduction to core concepts that students must understand in order to make independent contributions
- An extended research report giving more details and discussion than is possible in a conventional journal article
- A manual describing underlying principles and best practices for an experimental technique
- An essay exploring new ideas within physics, related philosophical issues, or broader topics such as science and society

Briefs allow authors to present their ideas and readers to absorb them with minimal time investment.

Briefs will be published as part of Springer's eBook collection, with millions of users worldwide. In addition, they will be available, just like other books, for individual print and electronic purchase.

Briefs are characterized by fast, global electronic dissemination, straightforward publishing agreements, easy-to-use manuscript preparation and formatting guidelines, and expedited production schedules. We aim for publication 8–12 weeks after acceptance.

More information about this series at http://www.springer.com/series/8902

Alexander Silbergleit · Arthur Chernin

Kepler Problem in the Presence of Dark Energy, and the Cosmic Local Flow

 Springer

Alexander Silbergleit
W.W. Hansen Experimental Physics
Laboratory
Stanford University
Stanford, CA, USA

Arthur Chernin
Sternberg Astronomical Institute
Moscow University
Moscow, Russia

ISSN 2191-5423 ISSN 2191-5431 (electronic)
SpringerBriefs in Physics
ISBN 978-3-030-36751-0 ISBN 978-3-030-36752-7 (eBook)
https://doi.org/10.1007/978-3-030-36752-7

This Springer imprint is published by the registered company Springer Nature Switzerland AG
The registered company address is: Gewerbestrasse 11, 6330 Cham, Switzerland

Preface

We consider the Newtonian motion of a particle in a gravitational field of a point mass immersed in the space filled with uniform dark energy (DE). As for any motion in the central field, the problem allows for the general solution in quadratures, which in this case can be expressed through elliptic integrals and/or elementary functions. We give all these solutions and examine them carefully.

Some significant results of our analysis are: (a) all purely radial infinite motions obey the Hubble law at large times; (b) all orbital (planar) infinite motions become asymptotically radial and obey the Hubble law at large times; (c) infinite planar motions strongly dominate the finite ones: the set of the energies, E, and orbital momenta, L, giving rise to the latter, is bounded, while the remaining unbounded part of the $\{E, L\}$ plane corresponds to former ones. This is clearly the effect of repulsive dark energy, since in the classical Kepler problem all orbital motions are finite for $E < 0$ and infinite in the opposite case. Another effect of DE is that bounded orbits are localized in space, predominantly within the equilibrium sphere, where the attractive Newtonian force balances exactly the repulsive force exerted by DE.

The main interest in this problem is, in the first place, due to recent investigation of the local flows of galaxies, which demonstrated that their dynamics is strongly dominated by dark energy, ensuring the Hubble law at distances of a few megaparsec. The local role of DE is intensively discussed in the book.

Brisbane, CA, USA
Moscow, Russia

Alexander Silbergleit
Arthur Chernin

Contents

Chapter 1
Introduction

In 1917 Einstein introduced the cosmological constant, Λ, to his equations of General Relativity, without a hint to its physical meaning. In 1922–1924, A. A. Friedman included the cosmological constant in his relativistic model of the universe and treated it as an empirical parameter to be measured through astronomical observations.

Much later, in 1965, Gliner [1] came up with the physical interpretation: Λ is proportional to the energy density of a special vacuum-like state of matter, which we now call dark energy. Based on this, after a few years Gliner produced [2] another fundamental idea, namely, of the initial vacuum-like state of the universe; such non-singular birth of the universe was discussed soon in paper [3].

These ideas, as well as the unique features of DE, were really tempting for physicists and cosmologists, who immediately got into a hot discussion of the new concept (see the authors' book [4]). Starting in the beginning of the 1980s, DE became increasingly popular resulting in the appearance and permanent growth of a new broad line of cosmological studies joined under the name of inflation, suggested by Guth [5], or blow-up of the universe, as Gliner [6] prefers to call it.

However, for a long time these activities looked more like some interesting, even if extravagant, theoretical insights, rather than anything close to reality. Until, around the turn of the century, well-known outstanding work on Supernovae Ia and anisotropy of cosmic microwave background revealed a fascinating fact: DE comprises almost 70% of our universe! This is the main result of the ΛCDM cosmological model, which is now universally accepted. Dark energy really governs our universe, being responsible, first of all, for the acceleration of its expansion detected in the supernovae observations.

But characteristic scale where the cosmological drama unveils is 300–1000 Mpc, at which distances the universe can be considered homogeneous, as required by the Friedmann cosmological model. So natural question is about the role, if any, of DE at much smaller scales of a few *Mpc*, as in Local Group and Flow of galaxies.

It appears that for the first time this question was raised, and some parts of theoretical answer were obtained, in [7–9]. It was found that the flow dynamics at 1–3 Mpc distances from the center of the Local Group is significantly dominated by dark energy. In particular, the Hubble law at these distances is directly due to

© The Author(s), under exclusive license to Springer Nature Switzerland AG 2019
A. Silbergleit and A. Chernin, *Kepler Problem in the Presence of Dark Energy, and the Cosmic Local Flow*, SpringerBriefs in Physics,
https://doi.org/10.1007/978-3-030-36752-7_1

the DE background. This implies that Hubble discovered actually local dark energy rather than the global cosmological expansion, which manifests itself at much larger distances mentioned above.

Further intensive work (see [10–20] and the references therein) involving an array of observational data confirmed, detailed, and extended the first results. A dozen of similar local flows around groups and clusters of galaxies were discovered [10, 13, 15, 18] at distances 10–30 Mpc, indicating that local galactic flows constitute a family of structures governed mostly by dynamical effects of dark energy.

We give a survey of all these results in Chap. 6 of the book. It is worthwhile to note a related but separate line of studies of effects on bounded systems produced by cosmological expansion and dark energy; an interested reader may be addressed to papers [21–28] and the references given there.

The influence of DE on the local cosmic flow is essentially a weak field effect (velocities $\ll c$, gravitational potential $\ll c^2$), which is why the Newtonian approximation is instrumental for studying it. A good model for the local group of galaxies is a massive ball on the uniform DE background creating a central field in which the galaxies in the local flow move as test particles. This alone demonstrates the fundamental role that the Kepler problem with DE plays in the local cosmological studies; of course, the two-body problem is the basis for any related investigations. So, although three- and many-body problems with DE were discussed in some cases [15, 18, 29], the main body of work was done using the Kepler problem in the presence of DE. It was studied analytically and numerically in [14, 16, 20]; in the following four chapters and Appendices A and B we give all its exact solutions and analyze them in detail, highlighting their significant properties. Chapter 7 contains our final comments and questions about dark energy.

Chapter 2
Two-Body Problem with Dark Energy

Abstract We formulate the problem of particle motion in the central field of a point mass and the uniform dark energy (DE) background. We introduce potential energy and describe the equilibrium (no-gravity) sphere. Equations of energy and angular momentum conservation are given showing two classes of motion, radial ones, with the zero angular momentum, and planar motions occurring when the angular momentum is non-zero. The classical Hubble law for all infinite motions holds due to DE presence.

2.1 Problem Set-Up

We consider a point mass M in a space filled with uniform DE of the density

$$\rho_\Lambda = \frac{c^2}{8\pi G} \Lambda \, ,$$

where Λ is the cosmological constant; the mass is at the origin of the coordinate system. We study all motions of a particle possible under such circumstances.

We deal with a particle of a unit mass, since the mass value drops from the gravitational equations of motion. On such a particle acts a Newtonian attracting force

$$\vec{F}_N = -\frac{GM}{r^2} \frac{\vec{r}}{r} \, , \tag{2.1}$$

and also an Einsteinian force from DE, $\vec{F}_E$, which turns out to be repulsive.

Indeed, by General Relativity the equation of state of DE is $p_\Lambda = -\rho_\Lambda c^2$, the effective gravitating density is

$$\rho_{eff} = \rho_\Lambda + 3p_\Lambda/c^2 = -2\rho_\Lambda, \tag{2.2}$$

A. Silbergleit and A. Chernin, *Kepler Problem in the Presence of Dark Energy, and the Cosmic Local Flow*, SpringerBriefs in Physics, https://doi.org/10.1007/978-3-030-36752-7_2

3

so the effective negative mass within the radius r is found to be

$$M_\Lambda(r) = \frac{4\pi}{3}\rho_{eff}r^3 = -\frac{8\pi}{3}\rho_\Lambda r^3 \ . \tag{2.3}$$

This results in a repulsive force

$$\vec{F}_E = -\frac{GM_\Lambda}{r^2}\frac{\vec{r}}{r} = \frac{8\pi}{3}(G\rho_\Lambda r)\frac{\vec{r}}{r} \equiv H^2 r \frac{\vec{r}}{r}\ , \tag{2.4}$$

where we introduced the Hubble parameter $H > 0$,

$$H^2 = \frac{8\pi}{3}G\rho_\Lambda = \frac{c^2\Lambda}{3}\ . \tag{2.5}$$

Based on this, the Kepler motion, $\vec{r} = \vec{r}(t)$, of a particle with DE present is described by the problem

$$\ddot{\vec{r}} = \vec{F}_N(\vec{r}) + \vec{F}_E(\vec{r}) = \left(-\frac{GM}{r^2} + H^2 r\right)\frac{\vec{r}}{r}\ ; \tag{2.6}$$

$$\vec{r}\,\Big|_{t=0} = \vec{r}_0, \qquad \dot{\vec{r}}\,\Big|_{t=0} = \vec{v}_0\ ;$$

here and elsewhere the dot over a symbol means the derivative in time t, and $\vec{r}_0$ and $\vec{v}_0$ are arbitrary initial position and velocity of the particle.

2.2 No-Gravity Sphere and Potential Energy

The total force on the right of the motion Eq. (2.6) vanishes when

$$r = r_\Lambda = \left(\frac{GM}{H^2}\right)^{1/3} = \left(\frac{3}{8\pi}\frac{M}{\rho_\Lambda}\right)^{1/3}\ . \tag{2.7}$$

So the 'no-gravity' sphere with the radius r_Λ consists of equilibria of the system (2.6). These rest points are unstable, as we will see in a moment. Remarkably, the amount of the repulsive (negative) mass inside the sphere is exactly equal to the attractive mass at its center, since, by the definitions (2.3) and (2.7),

$$-M_\Lambda(r_\Lambda) = \frac{8\pi}{3}\rho_\Lambda r_\Lambda^3 = M\ ; \tag{2.8}$$

hence the total gravitating mass inside r_Λ is exactly zero, which is why the gravity force vanishes at $r = r_\Lambda$.

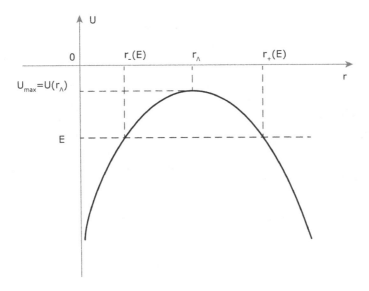

Fig. 2.1 Potential energy. Total energy $E < U_{max}$ is shown

We note the first significant difference between the classical Kepler problem and the one with DE: the latter has a characteristic distance, r_Λ, while there is no length scale in the former. It has deep consequences for the properties of solutions.

The equation of motion can be written in terms of the potential energy $U(r)$,

$$\ddot{\vec{r}} = -\frac{\partial U}{\partial \vec{r}}, \qquad U(r) = -\left(\frac{GM}{r} + \frac{1}{2}H^2 r^2\right) \equiv U_N(r) + U_E(r) . \qquad (2.9)$$

It is thus negative everywhere, and its only negative maximum is at $r = r_\Lambda$,

$$U_{max} = U(r_\Lambda) = -\frac{3}{2}\frac{GM}{r_\Lambda} = -\frac{3}{2}(GMH)^{2/3} . \qquad (2.10)$$

Therefore all the rest points at $r = r_\Lambda$ are indeed unstable, which is straightforward to find directly from the Eq. (2.6). The plot of $U(r)$ is given in Fig. 2.1.

2.3 Energy and Angular Momentum Conservation. The Hubble Law

The problem (2.6) is a particular case of the motion in the central field, so it allows for a solution in quadratures (e.g. [30], Sect. 14). It is obtained by means of two integrals of motion expressing conservation of energy and angular momentum:

$$\frac{1}{2}\dot{r}^2 + U(r) = E = \frac{1}{2}v_0^2 + U(r_0) = \text{const} ; \tag{2.11}$$

$$\vec{r} \times \dot{\vec{r}} = \vec{L} = \vec{r}_0 \times \vec{v}_0 = \text{const} . \tag{2.12}$$

Here $v_0 = |\vec{v}_0|$, and E and $\vec{L}$ are the total energy and angular momentum per unit mass; we drop 'per unit mass' in the sequel, saying just 'energy' and 'angular momentum'. We derive exact solution from Eqs. (2.11) to (2.12) in Chaps. 3, 4; right now we discuss some general features implied by the conservation laws.

Let us consider first an infinite motion, when a particle escapes to infinity, $r(t) \to \infty$ for $t \to \infty$. By the Eq. (2.11) and the definition (2.9) of the potential energy, to main order we have

$$\frac{1}{2}\dot{r}^2 - \frac{1}{2}H^2 r^2 = 0, \qquad r \to \infty ,$$

or

$$v = Hr, \quad t \to \infty; \qquad v(t) = |\dot{\vec{r}}(t)| . \tag{2.13}$$

This is the Hubble law: the relative velocity of two points is proportional to the distance between them; we examine it in more detail in Chaps. 3 and 4.

2.4 Radial and Orbital Motions

The angular momentum conservation shows two distinct types of motion.

2.4.1 Radial Motion: $\vec{v}_0 = v_0(\vec{r}/r)$, $L = 0$

In this case the Eq. (2.12) implies that the velocity of a particle is always parallel to its vector radius, $\dot{\vec{r}} \parallel \vec{r}$, which means that the particle moves all the way along the same radius. Eventually it either escapes to infinity, or falls on the central mass, as we demonstrate in Chap. 3.

2.4.2 Orbital (Planar) Motion: $\vec{v}_0 \neq v_0(\vec{r}/r)$, $L \neq 0$

By the Eq. (2.12), the particle moves in the plane perpendicular to the vector of the angular momentum, since $\vec{r} \cdot \vec{L} = 0$. The motion is either infinite, when the particle tends to infinity at large times, or finite; in the second case it orbits the central mass, but never falls on it, as shown in Chap. 4.

Chapter 3
Radial Motions: Exact Solution and Its Analysis

Abstract We describe the picture of radial motions depending on the value of the total energy, and give the exact solution to the problem in the form of an elliptic integral. The detailed form of the Hubble law is derived. Two solutions in terms of elementary functions are found, one of them coinciding with the Friedmann relativistic solution for expanding universe filled with no-pressure matter and DE. Other cases are reduced to a combination of the standard Legendre elliptical integrals and elementary functions, with derivations partly in Appendix A.

3.1 Finite and Infinite Motions: Fall on the Center and Escape to Infinity

The problem (2.6) for the radial motion $r = r(t)$ becomes

$$\ddot{r} = -\frac{GM}{r^2} + H^2 r \; ;$$

$$r\Big|_{t=0} = r_0, \qquad \dot{r}\Big|_{t=0} = v_0 \; ;$$

the energy conservation equation (2.11) turns to

$$\frac{1}{2}\dot{r}^2 + U(r) = E, \quad \text{or} \quad \frac{1}{2}\dot{r}^2 = E - U(r) \; . \tag{3.1}$$

Therefore the motion is only allowed where

$$E - U(r) > 0 \; ;$$

the available stretches depend thus on the total energy, E, determined by the initial position and velocity, r_0 and v_0.

If $E > U_{max} = U(r_\Lambda)$ [see formula (2.10)], then the motion is possible on the whole semi-axis $r > 0$; it is clearly seen in Fig. 2.1. If the initial velocity is negative,

© The Author(s), under exclusive license to Springer Nature Switzerland AG 2019
A. Silbergleit and A. Chernin, *Kepler Problem in the Presence of Dark Energy, and the Cosmic Local Flow*, SpringerBriefs in Physics,
https://doi.org/10.1007/978-3-030-36752-7_3

$v_0 < 0$, the particle moves towards the center, and reaches it in a finite time; the motion is finite. In the opposite case $v_0 > 0$ the particle goes away from the center and escapes to infinity at $t \to \infty$, so the motion is infinite.

When $E = U_{max} < 0$, finite and infinite motions are separated by the rest point $r = r_\Lambda$. If $r_0 < r_\Lambda$, then the particle falls on the center, bouncing first from r_Λ for $v_0 > 0$. It escapes to infinity when $r_0 > r_\Lambda$, bouncing first from r_Λ when $v_0 < 0$. As the equilibrium is unstable, no particle stays there in reality, being pushed away in one of the two directions by perturbations. If the perturbation velocity is not entirely radial, then the ensuing trajectory is no longer radial, as well.

Finally, in the case $E < U_{max} < 0$ the equation

$$E - U(r) = 0 \; ;$$

has two positive roots $r_\pm(E)$, $0 < r_- < r_\Lambda < r_+$, shown in Fig. 2.1. Here finite and infinite trajectories are separated by a whole segment $r_- < r < r_+$ forbidden for the motion. For $r_0 < r_-$ the particle falls on the center, bouncing first from r_- when $v_0 > 0$; for $r_0 > r_+$ it goes to infinity, bouncing first from r_+ if $v_0 < 0$. Apparently, $r_-(E) \to +0$, and $r_+(E) \to +\infty$ when $E \to -\infty$.

3.2   Exact Solution in Terms of the Radius r and Inverse Dimensionless Radius u. The Hubble Law

Expressing $\dot{r}$ from the energy conservation equation (3.1) and integrating leads to the exact solution of the radial motion problem,

$$t = \pm \int_{r_0}^{r} \frac{dr}{\sqrt{2\,[E - U(r)]}} = \pm \int_{r_0}^{r} \frac{dr}{\sqrt{2E + (2GM/r) + H^2 r^2}} \; , \qquad (3.2)$$

that gives the time as a function of the radius. The plus sign corresponds to the motion from the center, $\dot{r} > 0$, while the minus to the motion towards it, $\dot{r} < 0$.

Writing the above integrand explicitly as

$$\frac{r}{\sqrt{r\,(2Er + 2GM + H^2 r^3)}}$$

shows immediately that the integral is elliptic. However, it is more convenient to study it with the help of the dimensionless inverse radius

$$u = r_0/r \; ; \qquad (3.3)$$

note that $r = 0$ corresponds to $u = \infty$, $r = \infty$ corresponds to $u = 0$, and the initial value r_0 corresponds to $u_0 = 1$.

Changing the integration variable from r to u, after simple manipulations we find the solution giving time as a function of the dimensionless radius:

$$Ht = \pm \sqrt{b} \int_u^1 \frac{du}{u\sqrt{P(u)}}, \qquad P(u) = u^3 + au^2 + b, \qquad (3.4)$$

with the same meaning of the signs (recall that $u < 1$ for $r > r_0$, and $u > 1$ for $r < r_0$). The dimensionless coefficients of the cubic polynomial $P(u)$ are

$$a = \frac{Er_0}{GM} = \frac{E}{|U_N(r_0)|}; \qquad (3.5)$$

$$b = \frac{H^2 r_0^3}{2GM} = \frac{(8\pi/3)\rho_\Lambda r_0^3}{2M} = \frac{|M_\Lambda(r_\Lambda)|}{2M}\left(\frac{r_0}{r_\Lambda}\right)^3 = \frac{1}{2}\left(\frac{r_0}{r_\Lambda}\right)^3; \qquad (3.6)$$

to get the last expression for b, we made use of the formula (2.8).

We see that the parameter a is proportional to the total energy E, so it can be positive, zero, or negative together with it; in contrast with that, the parameter b is strictly positive, $b > 0$. Remarkably, we now have three sets of initial parameters, $\{r_0, v_0\}$, $\{r_0, E\}$, and $\{a, b\}$, of which only the last one is dimensionless. The l.h.s. of the equality (3.4) shows that the time scale of the motion is the inverse Hubble parameter, $1/H$.

In the next section we study the behavior of the key polynomial $P(u)$ in detail. Here we note just one its property needed for the Hubble law derivation, namely, that $P(0) = b$. Due to this we can write the following chain for the integral (3.4), aiming at the large time limit $u \to 0$ ($r \to \infty$) of an infinite motion:

$$\sqrt{b} \int_u^1 \frac{du}{u\sqrt{P(u)}} = \sqrt{b} \int_u^1 \frac{du}{u\sqrt{b}} + \sqrt{b} \int_u^1 \left[\frac{1}{u\sqrt{P(u)}} - \frac{1}{u\sqrt{b}}\right] du =$$

$$-\ln u + \int_u^1 \frac{\sqrt{b} - \sqrt{P(u)}}{u\sqrt{P(u)}} du = -\ln u + \int_u^1 \frac{b - P(u)}{u\sqrt{P(u)}\left[\sqrt{b} + \sqrt{P(u)}\right]} du =$$

$$-\ln u - \int_u^1 \frac{u^3 + au^2}{u\sqrt{P(u)}\left[\sqrt{b} + \sqrt{P(u)}\right]} du =$$

$$-\ln u - \int_0^1 \frac{au + u^2}{\sqrt{P(u)}\left[\sqrt{b} + \sqrt{P(u)}\right]} du + \int_0^u \frac{au + u^2}{\sqrt{P(u)}\left[\sqrt{b} + \sqrt{P(u)}\right]} du,$$

with the third term in the last line clearly vanishing as u^2 when $u \to 0$. Introducing the last expression to the solution (3.4) with the positive sign, turning from logs to exponents, and returning to the variable r produces

$$\frac{r}{r_0} = ke^{Ht}\left[1 + O\left(r_0^2/r^2\right)\right] = ke^{Ht}\left[1 + O\left(e^{-2Ht}\right)\right], \qquad t \to \infty, \qquad (3.7)$$

where

$$k = \exp \int_0^1 \frac{au + u^2}{\sqrt{P(u)}\left[\sqrt{b} + \sqrt{P(u)}\right]}\, du . \qquad (3.8)$$

This is the expression of the Hubble law for an infinite trajectory, complete with the proper coefficient depending on initial values, and a correction exponentially small at large times. The usual form (2.13) of the law, proportionality of the velocity to the distance, is obtained by differentiating the formula (3.7) in time.

For the sake of completeness, let us point out that, in contrast with the infinite time escape, a particle falls on the center, as stated above, in a final time

$$t_{fall} = -\frac{\sqrt{b}}{H} \int_\infty^1 \frac{du}{u\sqrt{P(u)}} ,$$

since the integral here apparently converges.

3.3   Properties of the Cubic Polynomial $P(u)$

According to its definition (3.4), the cubic polynomial to study is

$$P(u) = u^3 + au^2 + b, \qquad b > 0 .$$

As $P(0) = b > 0$ and $P(u) \to -\infty$ when $u \to -\infty$, it always has one negative root, which we denote $-\xi$, $\xi > 0$; when two other roots $u_{1,2}$ are real, they prove to be both positive, $0 < u_1 \le u_2$, as shown in Fig. 3.2. The two derivatives of the polynomial are

$$P'(u) = u(3u + 2a), \qquad P''(u) = 6u + 2a .$$

The first of them shows that $P(u)$ has two extrema, one at $u = 0$, the other at $u = -(2/3)a$. The second derivative at the extremum points are $P''(0) = 2a$, $P''(-(2/3)a) = -2a$. Hence at $u = 0$ there is a minimum for $a > 0$ and a maximum for $a < 0$, exactly opposite to the extremum at $u = -(2/3)a$, which is maximum for $a > 0$ and a minimum for $a < 0$; when $a = 0$, the two extrema coincide to form a deflection point at $u = 0$. Note also that

$$P(-(2/3)a) = (4/27)a^3 + b ,$$

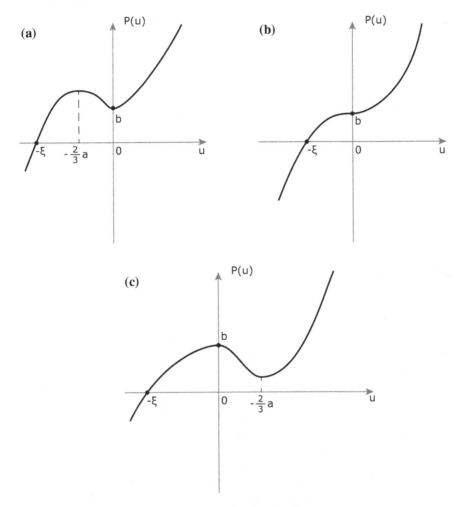

Fig. 3.1 $P(u)$ for $E > U_{max}$: **a** $E > 0$; **b** $E = 0$; **c** $0 > E > U_{max}$

which value is positive when

$$a > a_*, \quad a_* = -\frac{3}{2}(2b)^{1/3} = -\frac{3}{2}\frac{r_0}{r_\Lambda}, \quad \text{equivalent to} \tag{3.9}$$

$$E > -\frac{3}{2}\frac{GM}{r_\Lambda} = -\frac{3}{2}(GMH)^{2/3} = U_{max} \ .$$

Here we consequently used formulas (3.6), (3.5), (2.7), and (2.10). This is enough to plot $P(u)$ for $E > U_{max}$; the three plots corresponding to $E > 0$, $E = 0$, and $0 > E > U_{max}$ are given in Fig. 3.1.

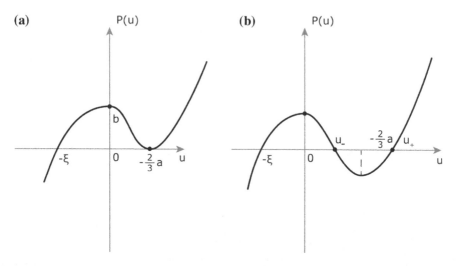

(a)   P(u)   **(b)**   P(u)

Fig. 3.2 $P(u)$ for $E \leq U_{max} < 0$: **a** $E = U_{max}$; **b** $E < U_{max}$

When a decreases further to and below a_* (E to and below U_{max}), two things happen. First, when $a = a_*$ ($E = U_{max}$), the minimum at $u = -3a_*/2 = r_0/r_\Lambda$ is zero, forming a double root $u_1 = u_2 = r_0/r_\Lambda$. For smaller a (smaller E) the minimum becomes negative, $u_1 = r_0/r_+$ drives towards zero, while $u_2 = r_0/r_-$ goes towards infinity. The radii $r_\pm(E)$ are defined in Sect. 3.1 as the roots of $E - U(r) = 0$, which is the same as $P(u) = 0$, in terms of the radius and inverse radius, respectively. The more negative is $a < a_*$, the deeper is the minimum, the closer u_1 to zero, and the larger u_2 going to infinity. However, a cannot, in fact tend to negative infinity for a fixed $b > 0$. Indeed, the initial position must always be in the zone $E - U(r) > 0$ where the motion is allowed, which is clearly equivalent to $P(1) = 1 + a + b > 0$. This is well understood in terms of the dimension initial data: $a \sim E r_0$, $b \sim r_0^3$, and when E gets more and more negative, the initial position for infinite motions $r_0 < r_-(E)$ tends to zero, while $r_0 > r_+(E)$ goes to infinity for the finite ones. The polynomial $P(u)$ for $E \leq U_{max}$ is plotted in Fig. 3.2.

Summarizing, the picture outlined here is, of course, in a one-to-one correspondence with findings of Sect. 3.1 based on the potential energy shown in Fig. 2.1: the first one just uses $u = r_0/r$ while the second deals with the radius r. Both the finite (fall on the center) and infinite (escape to infinity) motions exist for any value of the total energy. For $E > U_{max}$ both can start form any initial position, for $E = U_{max}$ they are separated by $u = r_0/r_\Lambda$, for $E < U_{max}$ the fall occurs for $r_0/r_- < u < \infty$, the escape for $0 < u < r_0/r_+$, with no motion possible in between.

3.4 Explicit Solutions in Elementary Functions

We turn to explicit radial solutions. As the escape solutions are mostly interesting for cosmological applications, we give only them here, i.e., calculate the integral (3.4) with the positive sign; the fall on the center solutions are obtained by reversing the sign of the corresponding expressions. In this section we present the solutions expressed by elementary functions for some special values of the initial parameters; there are just two such cases.

3.4.1 The Case E = 0

In this case $a = 0$, and the expression (3.4) turns to

$$Ht = \sqrt{b} \int_u^1 \frac{du}{u\sqrt{u^3 + b}} = \frac{2}{3} \int_1^{u^{-3/2}} \frac{d(\sqrt{b}s)}{\sqrt{1 + (\sqrt{b}s)^2}} =$$

$$\frac{2}{3} \left\{ \ln \left[b^{1/2}u^{-3/2} + \sqrt{1 + (b^{1/2}u^{-3/2})^2} \right] - \ln \left(\sqrt{b} + \sqrt{1+b} \right) \right\} ,$$

where we made the substitution $s = u^{-3/2}$. As $\ln(z + \sqrt{1 + z}) = \text{arcsinh } z$, we can rewrite this as

$$\frac{2}{3}Ht + \text{arcsinh}\sqrt{b} = \text{arcsinh} \left(b^{1/2}u^{-3/2} \right) ,$$

or, when inverted,

$$b^{1/2}u^{-3/2} = \sinh \left[\frac{2}{3}Ht + \text{arcsinh}\sqrt{b} \right] .$$

Turning from u to r and using the formula (3.6) for b, we arrive at the final result that holds for any initial position and $v_0 > 0$:

$$\frac{r}{r_\Lambda} = \left\{ 2^{1/2} \sinh \left[\frac{2}{3}Ht + \text{arcsinh} \left(\frac{r_0^3}{2r_\Lambda^3} \right)^{1/2} \right] \right\}^{2/3} ; \qquad (3.10)$$

naturally, it agrees with the general Hubble law (3.7) for $t \to \infty$.

This occurs to be the only of all solutions, both radial and orbital, where the radius is found as an explicit function of time, $r = r(t)$. Remarkably, the arbitrary initial position $r_0 > 0$ is involved in it only through the 'initial phase' of the hyperbolic sine, which ensures the initial condition $r(0) = r_0$.

3.4.2 The Case $E = U_{max} < 0$

Here $a = a_* = -(3/2)(r_0/r_\Lambda)$, and the polynomial $P(u)$ has a double root at $u = -(2/3)a_* = r_0/r_\Lambda$ (Fig. 3.2a); it is easy to find that the negative root is $-\xi = a_*/3 = -r_0/2r_\Lambda$. Therefore

$$P(u) = (u - a_*/3)(u + 2a_*/3)^2 ,$$

and, for $u < -(2/3)a_*$, the solution (3.4) becomes:

$$Ht = \sqrt{b} \int_u^1 \frac{du}{-u(u + 2a_*/3)\sqrt{u - a_*/3}} = -\frac{3\sqrt{b}}{2a_*} \int_u^1 \left(\frac{1}{u} - \frac{1}{u + 2a_*/3} \right) \frac{du}{\sqrt{u - a_*/3}} .$$

The two integrals on the utmost right is easily calculated (e.g. [31], **192.11**), so

$$Ht = \frac{\sqrt{b}}{2(-a_*/3)^{3/2}} \left[\ln \frac{\sqrt{u - a_*/3} - \sqrt{-a_*/3}}{\sqrt{u - a_*/3} + \sqrt{-a_*/3}} - \frac{1}{\sqrt{3}} \ln \frac{\sqrt{-a_*} - \sqrt{u - a_*/3}}{\sqrt{-a_*} + \sqrt{u - a_*/3}} \right] \Big|_u^1 =$$

$$\left[\ln \frac{u}{\left(\sqrt{u - a_*/3} + \sqrt{-a_*/3}\right)^2} - \frac{1}{\sqrt{3}} \ln \frac{\sqrt{-a_*} - \sqrt{u - a_*/3}}{\sqrt{-a_*} + \sqrt{u - a_*/3}} \right] \Big|_u^1 =$$

$$-\ln u - \ln \left[\left(\sqrt{u - a_*/3} + \sqrt{-a_*/3}\right)^2 \left(\frac{\sqrt{-a_*} - \sqrt{u - a_*/3}}{\sqrt{-a_*} + \sqrt{u - a_*/3}} \right)^{\frac{1}{\sqrt{3}}} \right] \Big|_u^1 ,$$

where we used the expression (3.9) for a_* to establish $\sqrt{b}/2(-a_*/3)^{3/2} = 1$. We now replace u with r_0/r in the first term of the last line, $(-a_*/3)$ with $(r_0/2r_\Lambda)$ everywhere, do the double substitution and take the exponent of both sides of the equality, to get the desired solution valid for $r_0 > r_\Lambda$ and $v_0 > 0$:

$$\frac{r}{r_0} = \frac{f(1)}{f(u)} e^{Ht}, \qquad 0 < u = \frac{r_0}{r} \leq 1 ; \tag{3.11}$$

$$f(u) = \left(\sqrt{u + r_0/2r_\Lambda} + \sqrt{r_0/2r_\Lambda} \right)^2 \left[\frac{\sqrt{3r_0/2r_\Lambda} - \sqrt{u + r_0/2r_\Lambda}}{\sqrt{3r_0/2r_\Lambda} + \sqrt{u + r_0/2r_\Lambda}} \right]^{\frac{1}{\sqrt{3}}} .$$

The answer clearly satisfies both the initial condition $r(0) = r_0$ and the Hubble law (3.7) at $t \to \infty$. Moreover, using the Taylor expansion of $f(u)$ around $u = 0$ and taking into account $f'(0) = 0$, it is not difficult to obtain the explicit dependence $r(t)$ in the form of a power series of the exponent $\exp(-Ht)$:

$$\frac{r(t)}{r_0} = \frac{f(1)}{f(0)} e^{Ht} + \sum_{n=1}^{\infty} d_n e^{-nHt}, \qquad d_1 = -\frac{f''(0)}{2f(0)} ,$$

with all other coefficients d_n determined from the proper polynomial recurrence relation $d_n = \mathcal{P}_n(d_1, d_2, \ldots, d_{n-1})$.

3.5 Explicit Solutions in Elliptic Integrals

As any elliptic integral, the solution (3.4) can be reduced to the combination of the standard elliptic integrals of by the known algorithm exploiting the rational fraction change of the integration variable (see [32], Sects. 13.3 and 13.5, or any other appropriate book). Here we choose to do it by simpler substitutions; they require some additional information about the polynomial $P(u)$.

First, since $P(-a) = b > 0$, the negative root $-\xi < -a$, or $\xi > a$, which is non-trivial for $a > 0$. Next, we can write

$$P(u) = (u + \xi)(u^2 - Au + \xi A) \equiv (u + \xi)Y(u), \qquad A = \xi - a > 0, \qquad (3.12)$$

which allows us to express two other roots of $P(u)$ through ξ:

$$u_{1,2} = \frac{1}{2}\left[A \pm \sqrt{A^2 - 4\xi A}\right] = \frac{1}{2}\left[(\xi - a) \pm \sqrt{-(\xi + a)(3\xi + a)}\right]. \qquad (3.13)$$

3.5.1 The Case $E > U_{max}$, $E \neq 0$

In this case $a > a_*$, and it is straightforward to see that $3\xi + a > 0$; so the roots (3.13) are complex, in an agreement with the Sect. 3.3 and Fig. 3.1. Thus the escape solution (3.4), by the equality (3.12), can be written as

$$Ht = \sqrt{b} \int_u^1 \frac{du}{u\sqrt{(u + \xi)(u^2 - Au + \xi A)}},$$

with the positive quadratic polynomial under the root. The reduction of this integral to a standard form is given in Appendix A; it results in the answer for the escape motions ($v_0 > 0$; note also that $r_0 > 0$ is arbitrary, and $0 < u \leq 1$):

$$Ht = 4\sqrt{\frac{b}{s_+ - s_-}} \times \qquad (3.14)$$

$$\left\{C_+\left[\Pi\left(x(u), \nu_+, k\right) - \Pi\left(x(1), \nu_+, k\right)\right] - C_-\left[\Pi\left(x(u), \nu_-, k\right) - \Pi\left(x(1), \nu_-, k\right)\right]\right\};$$

$$x(u) = \sqrt{1 - \frac{1}{s_+}\left[\sqrt{\left(u - \frac{\xi - a}{2}\right)^2 + \frac{(3\xi + a)(\xi - a)}{4}} - \left(u - \frac{\xi - a}{2}\right)\right]};$$

$$C_\pm = \frac{c_+}{(c_+ - c_-)(s_+ - c_\pm)}; \quad \nu_\pm = -\frac{s_+}{s_+ - c_\pm} < 0; \quad k^2 = \frac{s_+}{s_+ - s_-} < 1.$$

Here

$$\Pi(x, \nu, k) = \int\limits_{0}^{x} \frac{dx}{(1 + \nu x^2)\sqrt{(1 - x^2)(1 - k^2 x^2)}} \tag{3.15}$$

is the elliptic integral of the 3rd kind ([31], **772**), and $s_\pm$, $c_\pm$ are the roots of two quadratic polynomials that appear in the process of derivation. Their expressions through ξ and a are given in Eqs. (A.6) and (A.7), and their relevant properties are

$$s_- < 0 < s_+, \qquad c_- < 0 < c_+, \qquad s_+ - c_+ > 0 .$$

Recall also that $A = \xi - a > 0$ by the formula (3.12).

The result (3.14) apparently satisfies the initial condition $r(0) = r_0$, i.e., $u(0) = 1$. Its compliance with the asymptotic Hubble law (3.7) is by far not so obvious, but it is carefully checked at the end of the Appendix A.

3.5.2 The Case $E < U_{max} < 0$

In this case the polynomial $P(u)$ has, in addition to the negative root $(-\xi)$, two positive roots $u_1(\xi, a) < u_2(\xi, a)$ given in Eq. (3.13) [see Fig. 3.2b)]. Accordingly,

$$P(u) = (u + \xi)(u_1 - u)(u_2 - u) ,$$

so the escape ($v_0 > 0$) solution (3.4) becomes

$$Ht = \sqrt{b} \int\limits_{u}^{1} \frac{du}{u\sqrt{(u + \xi)(u_1 - u)(u_2 - u)}}, \qquad 0 < u \le 1 < u_1 .$$

Its reduction to standard elliptic integrals is pretty similar to the last step of the derivation carried out for the previous case in Appendix A, starting with the integrals (A.8), and thus much simpler.

We transform this integral to the form of Legendre by the substitution

$$\sqrt{u + \xi} = \sqrt{u_1 + \xi}\, x; \qquad x(u) = \sqrt{\frac{u + \xi}{u_1 + \xi}} , \tag{3.16}$$

which provides

$$Ht = -\frac{2}{\xi}\sqrt{\frac{b}{u_2 + \xi}} \int\limits_{x(u)}^{x(1)} \frac{dx}{(1 + \nu x^2)\sqrt{(1 - x^2)(1 - k^2 x^2)}} ,$$

with

$$\nu = -\frac{u_1 + \xi}{\xi} < 0; \qquad k^2 = \frac{u_1 + \xi}{u_2 + \xi} < 1 . \tag{3.17}$$

Using the definition (3.15) of the elliptic integral of the 3rd kind, it is straightforward to come up with the answer

$$Ht = \frac{2}{\xi} \sqrt{\frac{b}{u_2 + \xi}} \left[\Pi\left(x(u), \nu, k\right) - \Pi\left(x(1), \nu, k\right) \right] ; \tag{3.18}$$

all new parameters here are defined by the formulas (3.16) and (3.17). This result satisfies both the initial condition $r(0) = r_0$, with $r_0 \geq r_+$ (see Fig. 2.1), and the asymptotic Hubble law (3.7); we check the latter in the following way.

Noting $(\xi u_1 u_2)^{1/2} = b^{1/2}$, we find the leading singular term in the limit $u \to 0$,

$$Ht = \frac{2}{\xi} \sqrt{\frac{b}{u_2 + \xi}} \sqrt{\frac{(u_1 + \xi)(u_2 + \xi)}{u_1 u_2}} \int_0^{x(u)} \frac{dx}{1 - \left(\sqrt{\frac{u_1 + \xi}{\xi}} x\right)^2} =$$

$$2 \int_0^{x(u)} \frac{d\left(\sqrt{\frac{u_1 + \xi}{\xi}} x\right)}{1 - \left(\sqrt{\frac{u_1 + \xi}{\xi}} x\right)^2} = \int_0^{z(u)} \frac{dz}{1 - z} + \cdots = -\ln u + \cdots ,$$

where we set

$$\sqrt{\frac{u_1 + \xi}{\xi}} x = z, \qquad z(u) = \sqrt{\frac{u + \xi}{\xi}}$$

[see formula (3.16)]. Clearly, we obtained the leading term of the Hubble law.

This completes the explicit radial solutions describing the infinite motions (escape to infinity). There are four of them, (3.10), (3.11), (3.14), and (3.18); they cover the whole range of the total energy, $-\infty < E < \infty$.

3.6 Remark on Friedmann's Cosmology

In 1922 A. A. Friedmann [33] has established that the general relativistic scale factor, $a(t)$, of the cosmological expansion of pressureless matter in the presence of dark energy satisfies exactly the Eq. (3.1),

$$\dot{r}^2 = \frac{2GM}{r} + H^2 r^2 + E ,$$

if one replaces $r(t)$ with $a(t)$. The first term on the right is the contribution of matter, the second of DE, and E plays the role of the spacetime curvature: $E > 0$, $E = 0$, and $E < 0$ corresponds to the open, flat, and closed universe, respectively. So all explicit classical solutions found above are also general relativistic solutions describing the expansion of the ΛCDM universe.

In [33] Friedmann presented these solutions in the form of the integral (3.2), which was repeated later in his and many other articles. However, to the best of our knowledge, no explicit solutions were given during the next 80 years. In 2002 paper [34] was published with a number of explicit solutions describing the universe with DE and matter with any linear equation of state, $p = w\rho$, $w > -1$; all cases of solutions in terms of elliptic integrals are listed there. In particular, the solution for the flat universe with pressureless ($w = 0$) matter [formula (3.6) of the paper] coincides with our solution (3.10) for $E = 0$, if one sets $r_0 = 0$ in it.

Chapter 4
Orbital (Planar) Motions: Exact Solution and Its Analysis

Abstract We use conservation laws to show that the non-radial fall on the central mass is impossible, and that all infinite motions are asymptotically radial and obey the Hubble law. Its detailed form is given based on the solution in quadratures, which follows from the conservation equations. We study all the radial intervals where the motion is allowed depending on the total energy, E, and angular momentum, L. We demonstrate that finite motions are only possible for a finite interval of negative energies with the angular momentum below the critical value we found; so, unlike the classical Kepler problem, infinite motions grossly dominate the finite ones. We present and discuss all elementary function solutions existing for special values of E and L, such as circular and spiral orbits. All other cases lead to solutions in terms of Legendre elliptical integrals.

4.1 Conservation Laws. Inaccessible Center and Universal Escape

Let us return to the energy and angular momentum conservation laws (2.11), (2.12), now with $\vec{L} \neq 0$. We choose a Cartesian frame $\{x, y, z\}$ with the z axis along the angular momentum,

$$\vec{L} = L\hat{z} \; ;$$

the motion occurs thus in the $\{x, y\}$ plane, where we introduce polar coordinates $\{r, \phi\}$. From here on the vector radius $\vec{r}$ is two-dimensional, with the components

$$x = r \cos \phi, \qquad y = r \sin \phi \; .$$

In polar coordinates the angular momentum conservation becomes

$$r^2 \dot{\phi} = L = \text{const} \; ; \tag{4.1}$$

energy conserves according to

© The Author(s), under exclusive license to Springer Nature Switzerland AG 2019
A. Silbergleit and A. Chernin, *Kepler Problem in the Presence of Dark Energy, and the Cosmic Local Flow*, SpringerBriefs in Physics,
https://doi.org/10.1007/978-3-030-36752-7_4

$$\frac{1}{2}\left(\dot{r}^2 + r^2\dot{\phi}^2\right) + U(r) = E = \text{const} ,$$

where $U(r)$ is the potential energy (2.9). Expressing $\dot{\phi}$ from Eq. (4.1) we write:

$$\frac{1}{2}\left(\dot{r}^2 + \frac{L^2}{r^2}\right) + U(r) = E, \quad \text{or} \quad \frac{1}{2}\dot{r}^2 = E - U_e(r) ; \tag{4.2}$$

here we introduced the so called extended potential energy

$$U_e(r) = U(r) + \frac{L^2}{2r^2} = -\left(\frac{GM}{r} + \frac{1}{2}H^2r^2\right) + \frac{L^2}{2r^2} . \tag{4.3}$$

The radial stretches allowed for motion are now specified by

$$E - U_e(r) > 0 ; \tag{4.4}$$

however, the parameter behavior of $U_e(r)$ is significantly more complicated than that of $U(r)$, so we study the picture of possible motions, depending on the initial data, in terms of the dimensionless inverse radius $u = r_0/r$ in Sect. 4.3.

Two important facts, though, can be established immediately. One, the fall to the center is never possible ([30], Sect. 15): the condition (4.4) can be written as

$$Er^2 > -GMr - \frac{1}{2}H^2r^4 + \frac{L^2}{2} ,$$

which inequality cannot hold when $r \to 0$. Therefore in any finite motion a particle orbits the center in either a periodic or aperiodic way, but never reaches it. In contrast with that, infinity is always accessible due to the term $0.5H^2r^2$ leading $E - U_e(r)$ when $r \to \infty$ and making it positive at large radii, no matter what the total energy is.

Expressing $\dot{r}$ from Eq. (4.2) and integrating provides

$$t = \pm \int_{r_0}^{r} \frac{dr}{\sqrt{2\left[E - U_e(r)\right]}} = \pm \int_{r_0}^{r} \frac{dr}{\sqrt{2E + (2GM/r) + H^2r^2 - \left(L^2/r^2\right)}} .$$

Next, by writing $\dot{\phi} = (d\phi/dr)\dot{r}$, from this and Eq. (4.1) we derive

$$\phi - \phi_0 = \pm \int_{r_0}^{r} \frac{L\, dr}{r^2\sqrt{2E + (2GM/r) + H^2r^2 - \left(L^2/r^2\right)}} , \tag{4.5}$$

where ϕ_0 is the initial phase. So we have the solution in the form of integrals, which are apparently elliptic. As before, the plus sign corresponds to the motion outwards, $\dot{r} > 0$, the minus to the inward motion, $\dot{r} < 0$.

4.2 Exact Solution in Terms of the Inverse Radius u. Infinite Motions and the Hubble Law

Turning to $u = r_0/r$ in the above integrals, we obtain the solution in the form

$$
Ht = \pm\sqrt{b}\int\limits_{u}^{1}\frac{du}{u\sqrt{Q(u)}}, \qquad \phi - \phi_0 = \pm\sqrt{l}\int\limits_{u}^{1}\frac{u\,du}{\sqrt{Q(u)}}, \tag{4.6}
$$

where we introduced a polynomial of the 4th degree proportional to $E - U_e(r)$:

$$
Q(u) = -lu^4 + u^3 + au^2 + b . \tag{4.7}
$$

Its coefficients a and b are defined by the formulas (3.5) and (3.6), while l is a new dimensionless parameter proportional to the square of the angular momentum,

$$
l = \frac{L^2}{r_0^2}\frac{r_0}{2GM} = \frac{L^2}{2GMr_0} > 0 . \tag{4.8}
$$

Thus we reduced, due to the choice of frame, the original set $\{\vec{r}_0, \vec{v}_0\}$ of six scalar initial values first to $\{r_0, \phi_0, E, L\}$, and then to dimensionless $\{\phi_0, a, b, l\}$, with r_0 involved in the last three parameters. In addition, the sign of the initial radial velocity is reflected in plus/minus signs in the formulas. The values of the initial velocity components are

$$
\dot{r}(0) = \pm\sqrt{2\,[E - U_e(r_0)]}, \qquad r_0\dot{\phi}(0) = L/r_0 ;
$$

furthermore E, L, and r_0 can be replaced with a, l and b according to the formulas (3.5), (4.8), and (3.6).

It is worthwhile to note that when there is no DE, i.e., $\rho_\Lambda = 0$, $H = 0$, both the radial and orbital solutions represented by integrals in r obviously turn to the classical Kepler solutions. It is less obvious when the variable $u = r_0/r$ is used in the formulas (3.4) and (4.6); the transition still holds, of course, since, by the definition (3.6), $b = 0$ in this case, and always $\sqrt{b}/H = \sqrt{r_0^3/2GM}$.

The intervals allowed for motion are now specified by $Q(u) > 0$; as $u(0) = 1$,

$$
Q(1) = -l + 1 + a + b > 0 .
$$

For any infinite motion, i.e., when $r(t) \to 0$ at $t \to \infty$, the angular momentum conservation (4.1) implies $\dot{\phi}(t) \to 0$, so the phase tends to a constant, $\phi(t) \to \phi_\infty$. Thus all infinite motions are asymptotically radial. The same chain of transformations of the first integral (4.6) as that for a radial solution presented in Sect. 3.2, leads to the same form of the Hubble law as in Eq. (3.7):

$$\frac{r}{r_0} = ke^{Ht}\left[1 + O\left(r_0^2/r^2\right)\right] = ke^{Ht}\left[1 + O\left(e^{-2Ht}\right)\right], \qquad t \to \infty ; \qquad (4.9)$$

$$k = \exp \int_0^1 \frac{au + u^2 - lu^3}{\sqrt{Q(u)}\left[\sqrt{b} + \sqrt{Q(u)}\right]} \, du ;$$

the difference is only in the value of the coefficient k, due to terms with the parameter l. By differentiating first Eq. (4.9) in time one obtains

$$v = Hr, \qquad r \to \infty ,$$

which is exactly the classical form (2.13) of the Hubble law.

4.3 The Key 4th Degree Polynomial $Q(u)$

4.3.1 Behavior of $Q(u)$ Under Changing Initial Parameters

It is convenient to study the changes in the behavior of $Q(u)$ depending on the total energy, or parameter a, growing from negative to positive values, and not decreasing, as in Sect. 3.3 where $P(u)$ was investigated.

By the definition (4.7), $Q(0) = b > 0$ and $Q(u) \to -\infty$ for $u \to \pm\infty$, so the polynomial always has a negative root, which we again denote $-\xi$, $\xi > 0$, and a positive one denoted $\eta > 0$: $Q(-\xi) = Q(\eta) = 0$. As we show below, sometimes $Q(u)$ has two more positive roots $u_{1,2}$; we always keep the notation $\eta > 0$ for *the smallest positive root*, so when all the roots are real we have

$$-\xi < 0 < \eta \le u_1 \le u_2 .$$

We will need the derivatives of the polynomial,

$$Q'(u) = u(-4lu^2 + 3u + 2a), \qquad Q''(u) = 2(-6lu^2 + 3u + 2a) . \qquad (4.10)$$

We see that $Q'(0) = 0$, $Q''(0) = 4a$, so $u = 0$ is always an extremum of $Q(u)$: it is maximum for $a < 0$, a deflection point at $a = 0$, and minimum when $a > 0$. The other two roots of the first derivative,

Fig. 4.1 $Q(u)$ for $a < a^* < 0$

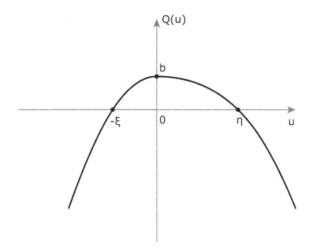

$$u_\pm = \frac{1}{8l}(3 \pm \sqrt{9 + 32al}), \qquad Q'(u_\pm) = 0, \tag{4.11}$$

are complex conjugate when

$$a < a^* \equiv -\frac{9}{32\,l} < 0, \quad \text{or} \tag{4.12}$$

$$E < E^* = a^* \frac{GM}{r_0} = -\frac{9}{32\,l} \frac{GM}{r_0} = -\frac{9}{16}\left(\frac{GM}{L}\right)^2 < 0;$$

we used the definition (4.8) of parameter l to get the last expression here.

Thus for $a < a^*$ the polynomial has no other extrema; the corresponding plot is shown in Fig. 4.1. In fact, the plot of $Q(u)$ looks precisely as in Fig. 4.1 when $a < -(9/24\,l) < a^*$. For larger values of a the second derivative $Q''(u)$ has two real roots, at which points the plot of $Q(u)$ changes its convexity, from concave to convex, or vice versa. As these details do not play any role in our investigation, and to avoid extra figures, we do not show those ripples in our plots. When $a \geq a^*$, the polynomial $Q(u)$ has two more extrema at $u_\pm$ that significantly affect the appearance of its plot. At $a = a^*$ these roots of the derivative coincide,

$$u_+(a^*) = u_-(a^*) = \frac{3}{8l} > 0,$$

forming the deflection point of $Q(u)$. Than they separate, with the smaller one apparently moving all the way to the left, and the larger to the right.

The evolution following the crossing of a^* by a depends, however, also on the parameter b, since the value of the polynomial at the critical point depends on it:

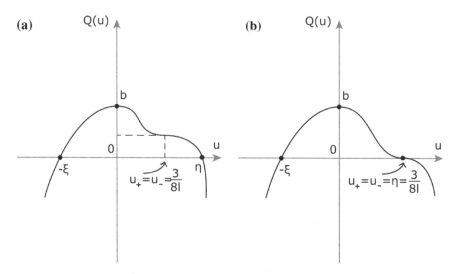

Fig. 4.2 $Q(u)$ for $a = a_*$: **a** $b > b_*$; **b** $b = b_*$

$$Q(u_\pm)\Big|_{a=a^*} = Q\left(\frac{3}{8l}\right)\Big|_{a=a^*} = b - b_*, \qquad b_* \equiv \left(\frac{3}{16l}\right)^3 > 0. \qquad (4.13)$$

So for $b > b_*$ the value of the polynomial at the deflection point is positive, and $\eta > 3/8l$, Fig. 4.2a; for $b = b_*$ we have $Q(3/8l) = 0$, $\eta = 3/8l$, Fig. 4.2b. We note that $Q(u_\pm)$ always grows with a unless $u_- = 0$ (when $a = 0$), because

$$\frac{dQ(u_\pm)}{da} = Q'(u_\pm)\frac{du_\pm}{da} + u_\pm^2 = u_\pm^2 > 0. \qquad (4.14)$$

This helps to understand the structural changes in $Q(u)$ for $a > a_*$ and $b \geq b_*$. When $a_* < a < 0$, our polynomial has a maximum at $u = 0$, a minimum at $u = u_- > 0$, and another maximum at $u = u_+$ (all the extrema positive, Fig. 4.3). At $a = 0$, also $u_- = 0$, hence $u = 0$ becomes a deflection point, with $u_+ = 3/4l$ remaining the only maximum, Fig. 4.4a; in both cases $\eta > u_+$. Finally, when a becomes positive, u_- turns negative and is a maximum of $Q(u)$, its minimum is at $u = 0$, and the second maximum is at $u = u_+ < \eta$, as shown in Fig. 4.4b.

In the remaining case $0 < b < b_*$ several significant events happen when a varies from $a_* < 0$ to zero. The value $Q(u_\pm)$ is now negative at $a = a^*$, Fig. 4.5a. By the inequality (4.14), for slightly larger a the plot must look as in Fig. 4.5b, with the negative maximum at $u = u_+$ that keeps growing with a.

It is straightforward to calculate that $Q(u_+)\Big|_{a=0} > 0$, therefore, by the Cauchy theorem, it turns to zero at some $a = a'$ between a^* and $a = 0$,

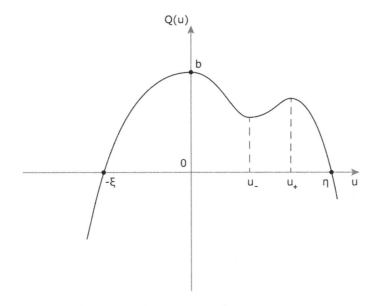

Fig. 4.3 $Q(u)$ for $b \geq b_*$, $a^* < a < 0$

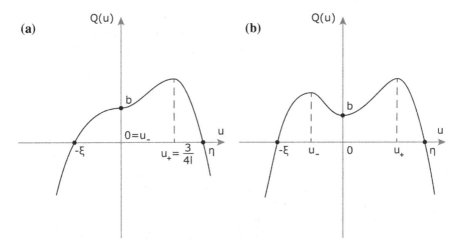

Fig. 4.4 $Q(u)$ for $b \geq b_*$: **a** $a = 0$; **b** $a > 0$

$$Q(u_+)\Big|_{a=a'} = 0, \qquad a^* < a' < 0 , \tag{4.15}$$

which, in fact, is the definition of a'. The corresponding plot is given in Fig. 4.6a. The polynomial has a zero maximum and a double root at $u = u_+ = u_1 = u_2$, a negative minimum at $u = u_-$, and, of course, a positive maximum at $u = 0$, with $\eta < u_-$.

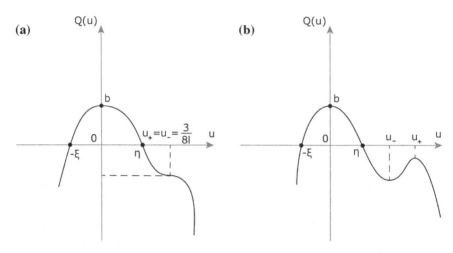

Fig. 4.5 $Q(u)$ for $b < b_*$: **a** $a = a^* < 0$; **b** $a^* < a < a' < 0$

For negative values of a slightly larger than a' the maximum at $u = u_+$ becomes positive, the roots $u_{1,2}$ separate, $u_+ < u_1 < u_2$, and the second segment allowed for motion, $u_1 < u < u_2$, $Q(u) > 0$, appears between them, Fig. 4.6b.

The negative minimum at $u = u_-$, however, keeps growing with a according to the inequality (4.14), and $Q(u_-)\big|_{a=0} = Q(0) = b > 0$. Applying the Cauchy theorem again, we see that at some $a = a''$ between a' and zero, $Q(u_-)$ must be zero,

$$Q(u_-)\bigg|_{a=a''} = 0 \qquad a^* < a' < a'' < 0 , \tag{4.16}$$

which requirement serves as a definition of a''. For $a = a''$, $0 < b < b_*$ the graph of the polynomial $Q(u)$ is presented in Fig. 4.6c. Now it has a positive maximum at $u = u_+$, $u_1 < u_+ < u_2$, a zero minimum at $u = u_- = u_1 = \eta$, and the multiply mentioned positive maximum $Q(0) = b$.

For $a'' < a < 0$, $0 < b < b_*$ the minimum becomes positive, the plot is that of Fig. 4.3; then at $a = 0$ it becomes as in Fig. 4.4a, and finally, for positive values of a, i.e., of the total energy, as in Fig. 4.4b.

Here is the summary of the evolution of $Q(u)$ with a growing from minus to plus infinity as established above: for $b \geq b_* > 0$ it goes according to the Figs. 4.1, 4.2a or b, 4.3, 4.4a, b; for $0 < b < b_*$ the proper sequence of figures is 4.1, 4.5a, b, 4.6a–c, 4.3, 4.4a, b.

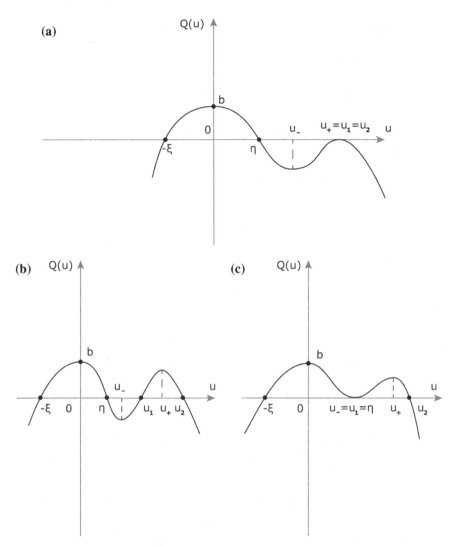

Fig. 4.6 $Q(u)$ for $b < b_*$: **a** $a = a' < 0$; **b** $a' < a < a'' < 0$; **c** $a = a'' < 0$

4.3.2 Some Results Related to the Behavior of $Q(u)$

The definitions (4.15) and (4.16) do not provide a convenient analytical expressions for the critical values a' and a'', because they reduce to an algebraic equation of the 4th degree for a whose proper roots are the sought values of a' and a''. This calculation is still significant showing that they scale with the parameter l as

$$a' = -\alpha'/l, \qquad a'' = -\alpha''/l, \qquad 0 < \alpha'' < \alpha' < |a^*l| = 9/32,$$

with alphas depending only on a small positive number $\beta = bl^3$: due to $b < b_*$, $0 < \beta < (3/16)^3$. The corresponding asymptotic calculations of Appendix B lead to the following expansions in this small parameter:

$$\alpha' = \frac{1}{4} + 4\beta + \cdots, \qquad \alpha'' = \frac{3}{2}(2\beta)^{1/3} - (2\beta)^{2/3} + \cdots ; \qquad (4.17)$$

we see that α' is larger than one quarter, tending to it when $\beta \to 0$, while α'' is closer to zero and goes to it in this limit.

In line with the notation (4.12) for E^*, we denote the critical values of energy and angular momentum in the following way:

$$L_*^2 = \frac{3}{8b_*^{1/3}} GMr_0 = \frac{3}{2^{8/3}} GMr_\Lambda = \frac{3}{2^{8/3}} \left[\frac{(GM)^2}{H}\right]^{2/3} ;$$

$$E' = a' \frac{GM}{r_0} = -\frac{\alpha'}{l} \frac{GM}{r_0} = -2\alpha' \left(\frac{GM}{L}\right)^2 = -\alpha' \frac{2^{11/3}}{3} \left(\frac{L_*}{L}\right)^2 \frac{GM}{r_\Lambda} ; (4.18)$$

$$E'' = a'' \frac{GM}{r_0} = -\frac{\alpha''}{l} \frac{GM}{r_0} = -2\alpha'' \left(\frac{GM}{L}\right)^2 = -\alpha'' \frac{2^{11/3}}{3} \left(\frac{L_*}{L}\right)^2 \frac{GM}{r_\Lambda} ;$$

take notice that $b < b_*$ corresponds to $L < L_*$, etc.. Note also that the new notation L_* allows for a useful alternative expression for parameter $\beta = bl^3$:

$$\beta = \left(\frac{3}{16}\right)^3 \left(\frac{L}{L_*}\right)^6 , \qquad \beta^{1/3} = \frac{3}{16} \left(\frac{L}{L_*}\right)^2 . \qquad (4.19)$$

Combined with expressions (4.18) for the critical energies and with the asymptotics (4.17), this provides the effective values of E' and E'' to l.o. in β, or L/L_*:

$$E' \approx -\frac{2^{5/3}}{3} \left(\frac{L_*}{L}\right)^2 \frac{GM}{r_\Lambda}, \qquad E'' \approx -\frac{3}{2} \frac{GM}{r_\Lambda} . \qquad (4.20)$$

Remarkably, the value of E'' to l.o. does not depend on the initial parameters and coincides with the maximum potential energy (2.10).

Moreover, in elliptic integral solutions obtained in Sect. 4.6 the real roots $u_{1,2}$ of $Q(u)$ are involved, along with $-\xi$ and η existing always,

$$-\xi < 0 < \eta \leq u_1 \leq u_2 .$$

To reduce the number of parameters in the answers we give here the expressions of $u_{1,2}$ through ξ and η. We have

$$Q(u) = -lu^4 + u^3 + au^2 + b = -l(u + \xi)(u - \eta)(u - u_1)(u - u_2) ,$$

so, as in Viet theorem, by comparing the coefficients at u^3 and u^0 on both sides, we obtain:

$$\xi - \eta - u_1 - u_2 = -1/l, \qquad b = l\xi\eta \, u_1 u_2 , \qquad (4.21)$$

or

$$u_1 + u_2 = 1/l + \xi - \eta, \qquad u_1 u_2 = b/l\xi\eta . \qquad (4.22)$$

Hence $u_{1,2}$ are the roots of the quadratic equation

$$z^2 - (1/l + \xi - \eta)\, z + b/l\xi\eta = 0 ,$$

and their desired expressions are:

$$u_{1,2} = \frac{1}{2}\left[\left(\frac{1}{l} + \xi - \eta\right) \pm \sqrt{(1/l + \xi - \eta)^2 - \frac{4bl}{\xi\eta}} \right] ; \qquad (4.23)$$

the minus sign corresponds to the root u_1, the plus sign to u_2, and both roots are positive.

We also mark one inequality, which follows from another Viet identity [coefficients at u in the two representations of $Q(u)$],

$$\xi(u_1 u_2 + \eta u_1 + \eta u_2) - \eta u_1 u_2 = 0, \quad \text{or} \quad u_1 u_2(\xi - \eta) + (u_1 + u_2)\xi\eta = 0 .$$

Using here the identities (4.22), we find

$$b = l(\xi\eta)^2 \, \frac{1/l + \xi - \eta}{\eta - \xi} .$$

As $b > 0$ and $l > 0$, this implies the inequality

$$0 < \xi < \eta < \xi + 1/l . \qquad (4.24)$$

Thus the distance between zero and η is always larger than that between $-\xi$ and zero, but they become equal in the limit of large angular momentum, $l \to \infty$. In particular, the quatric parabola in Fig. 4.1 becomes symmetric in this limit.

4.4 Dominating Infinite, and Finite Motions

The above analysis demonstrates that infinite planar motions are always possible, for the whole range of initial parameters $-\infty < E < \infty$, $0 < L \neq 0 < \infty$, provided that the starting radius r_0 is large enough. If the initial radial velocity is negative, then the particle reaches $r_{min} = r_0/\eta$ in a finite time, turns to moving towards larger

radii, and eventually escapes to infinity, as in the case of positive radial velocity at
the start of motion; two deviations from such behavior, for certain values of E and
L, are indicated in the next section. As shown in Sect. 4.2, all infinite motions are
asymptotically radial and obey the Hubble law.

In contrast with that, finite orbital motions in the interval
$0 < u_1 < u < u_2$, or $0 < r_0/u_2 < r < r_0/u_1$, exist in a bounded range of energy,
$E' < E < E''$, and angular momentum, $0 < L < L_*$, as seen in Fig. 4.6b, c. For
$L = 0$ there is also a radial fall on the center; a few more finite orbits generated by
special initial parameters are described in Sect. 4.5. All those special solutions still
occur at $0 < L \leq L_*$, so L_* is really a threshold value of the angular momentum: no
finite motions are possible above it.

Clearly, the domination of infinite planar motions over finite ones is the effect
of repulsive dark energy, because in the classical Kepler problem all motions with
non-negative energy $E \geq 0$ are infinite (with the exception of the radial fall on the
center), while those with the negative energy $E < 0$ are finite.

A particle on a generic finite trajectory oscillates in the radial direction between
$r_{min} = r_0/u_2$ and $r_{max} = r_0/u_1$ according to the formulas (4.6) with the plus or minus
sign, which changes to the opposite at every bounce from the radial turning points.
The orbit is periodic under the condition

$$\frac{2\pi m}{n} = 2\sqrt{l} \int_{u_1}^{u_2} \frac{u\,du}{\sqrt{Q(u)}}, \qquad m,\ n \text{ integer}, \qquad (4.25)$$

in which case, after n trips from r_{min} to r_{max} and back, the particle returns to the
original point characterized by the polar angle $\phi \pmod{2\pi}$. If the condition (4.25)
holds, then the period of this motion is

$$T = 2n \frac{\sqrt{b}}{H} \int_{u_1}^{u_2} \frac{du}{u\sqrt{Q(u)}} = n\sqrt{\frac{2r_0^3}{GM}} \int_{u_1}^{u_2} \frac{du}{u\sqrt{Q(u)}}, \qquad m,\ n \text{ integer}, \qquad (4.26)$$

where H is the Hubble parameter (2.5).

4.5   Explicit Orbital Solutions in Elementary Functions

4.5.1   Circular Orbits

The motion along a circle $r = r_0 = \text{const}$ is not described by the expressions (4.6),
because integration over r is not possible in this (and only this) case. Apparently,
such trajectory is given by

$$r = r_0 = \text{const}, \qquad \phi - \phi_0 = \omega t, \qquad \omega = \sqrt{\frac{GM}{r_0^3} - H^2}. \qquad (4.27)$$

So DE reduces the Kepler orbital frequency, and circular motion can take place only inside the equilibrium sphere,

$$r_0 < \left(\frac{GM}{H^2}\right)^{1/3} = r_\Lambda.$$

Combining the above formula for $\omega = \dot{\phi}$ with the equation of the angular momentum conservation (4.1) we obtain

$$L^2 = GMr_0 - H^2 r_0^4 = GMr_0 \left[1 - \left(\frac{r_0}{r_\Lambda}\right)^3\right]. \qquad (4.28)$$

Expression (4.28) as a function of r_0 on the interval $0 < r_0 < r_\Lambda$ has a single maximum at $r_0 = r_\Lambda/4^{1/3}$. Due (2.7) and (4.18), this leads to a remarkable result: the orbital momentum of any circular orbit does not exceed L_*,

$$L^2 \le GMr_\Lambda \frac{1}{4^{1/3}} \left(1 - \frac{1}{4}\right) = \frac{3}{2^{8/3}} GMr_\Lambda = L_*^2, \quad \text{or} \quad 0 < L \le L_*. \qquad (4.29)$$

On the other hand, from the same equality (4.28) one immediately gets a lower L-dependent bound for the orbit radius:

$$r_0 > \frac{L^2}{GM} = \frac{L^2}{GMr_\Lambda} r_\Lambda = \frac{3}{2^{8/3}} \left(\frac{L}{L_*}\right)^2 r_\Lambda,$$

so

$$\frac{3}{2^{8/3}} \left(\frac{L}{L_*}\right)^2 r_\Lambda < r_0 < r_\Lambda. \qquad (4.30)$$

The estimates (4.29) and (4.30) are in a stark contrast with the classical Kepler case where neither the orbit radius nor the orbital momentum is bounded.

Next, the energy conservation equation along with the Eqs. (4.28) and (2.7) provides the value of the total energy and its upper bound for a circular orbit:

$$E = U_e(r_0) = -\left(\frac{GM}{r_0} + \frac{1}{2} H^2 r_0^2\right) + \frac{L^2}{2r_0^2} = -\left(\frac{GM}{2r_0} + H^2 r_0^2\right) \le -\frac{3}{2^{2/3}} \frac{GM}{r_\Lambda}. \qquad (4.31)$$

Thus any orbit radius $r_0 < r_\Lambda$ uniquely specifies the angular velocity, angular momentum, and total energy. The same is true if, instead, the value of the angular frequency $\omega > 0$ is given: one first finds the unique value of the orbit radius by the formula (4.27), and then the following L and E:

$$r_0 = \frac{r_\Lambda}{(1+\omega^2/H^2)^{1/3}}; \quad L = \frac{r_\Lambda^2\,\omega}{(1+\omega^2/H^2)^{2/3}}; \quad E = -\frac{GM}{r_\Lambda}\frac{2+\omega^2/H^2}{(1+\omega^2/H^2)^{2/3}}\,.$$

$$\text{(4.32)}$$

Interestingly, the maximum value of the orbital momentum $L = L_*$ is achieved when $\omega = \sqrt{3}H$ and $r_0 = 2^{-2/3}r_\Lambda$.

Finally, for given E and L satisfying, respectively, conditions (4.29) and (4.31), first the orbit radius is determined uniquely from the energy and momentum conservation, and then a unique value of ω is found. The expressions are cumbersome, so we give just the first of them:

$$r_0 = \frac{GM(2H^2L^2+E^2)}{4(GM)^2H^2+E(H^2L^2+E^2)} = r_\Lambda\frac{E_\Lambda(8E_HE_L+E^2)}{8E_\Lambda^2E_H-|E|(4E_HE_L+E^2)}\,;$$

to clarify the dimensions, we introduced three positive characteristic energies,

$$E_\Lambda = \frac{GM}{r_\Lambda}, \quad E_H = \frac{H^2r_\Lambda^2}{2}, \quad E_L = \frac{L^2}{2r_\Lambda^2}\,;$$

we also took into account that the total energy is negative, $E = -|E|$.

Other exact elementary orbital solutions occur when the polynomial $Q(u)$ has a multiple, double or triple, root. All such cases are shown in Figs. 4.2b, 4.6a, c; we cover them one by one.

4.5.2 The Case $E = E^*$, $L = L_*$

This case, with $Q(u)$ plotted in Fig. 4.2b, is described by dimensionless parameters $a = a_* = -9/32l$ and $b = b_* = (3/16l)^3$. The polynomial has a triple root $u_+ = u_- = \eta = 3/8l$; it is straightforward to see that $\xi = 1/8l = (1/3)\eta$, so the polynomial becomes $Q(u) = l(\eta/3+u)(\eta-u)^3$. Moreover, here

$$b = \frac{1}{2}\left(\frac{r_0}{r_\Lambda}\right)^3 = b_* = \left(\frac{3}{16l}\right)^3, \text{ hence } \frac{1}{l} = \frac{2^{11/3}}{3}\frac{r_0}{r_\Lambda}, \text{ and } \eta = \frac{3}{8l} = 2^{2/3}\frac{r_0}{r_\Lambda}\,.$$

The integrals (4.6) are calculated easily (c.f. [31], **380.111, 380.001**):

$$\sqrt{b}\int_u^1\frac{du}{u\sqrt{Q(u)}} = \sqrt{\frac{b}{l}}\int_u^1\frac{du}{u(\eta-u)\sqrt{(\eta/3+u)(\eta-u)}} =$$

$$\frac{1}{\eta}\sqrt{\frac{b}{l}}\left[\int_u^1\frac{du}{u\sqrt{(\eta/3+u)(\eta-u)}} + \int_u^1\frac{du}{(\eta-u)\sqrt{(\eta/3+u)(\eta-u)}}\right] =$$

$$-\ln u + \ln\frac{\sqrt{(\eta+3u)(\eta-u)}+u+\eta}{\sqrt{(\eta+3)(\eta-1)}+1+\eta} + \sqrt{\frac{\eta+3}{\eta-1}} - \sqrt{\frac{\eta+3u}{\eta-u}}\,;$$

$$\sqrt{l}\int_u^1 \frac{u\,du}{\sqrt{Q(u)}} = \int_u^1 \frac{u\,du}{(\eta - u)\sqrt{(\eta/3 + u)(\eta - u)}} =$$

$$-\int_u^1 \frac{du}{\sqrt{(\eta/3 + u)(\eta - u)}} + \eta\int_u^1 \frac{du}{(\eta - u)\sqrt{(\eta/3 + u)(\eta - u)}} =$$

$$\sqrt{3}\left(\sqrt{\frac{\eta + 3}{\eta - 1}} - \sqrt{\frac{\eta + 3u}{\eta - u}}\right) + \left(\arcsin\frac{\eta - 3}{2\eta} - \arcsin\frac{\eta - 3u}{2\eta}\right) \quad ;$$

for the first one we used the second identity (4.22) in the form $(1/\eta^2)\sqrt{3b/l} = 1$.

These results work, in fact, for two different solutions, both with the same special values of the angular momentum and total energy; by formulas (4.18) and (4.12), those are

$$L = L_* = \frac{3^{1/2}}{2^{4/3}}(GMr_\Lambda)^{1/2}, \qquad E = E^* = -\frac{9}{16}\left(\frac{GM}{L_*}\right)^2 = -\frac{3}{2^{4/3}}\frac{GM}{r_\Lambda}.$$
$$(4.33)$$

One solution describes an infinite motion with the positive initial radial velocity, $\dot{r}(0) > 0$, and, accordingly, the positive signs in the formulas (4.6). For $u \to 0$, or $r \to \infty$, the leading term in the first integral expression above is $\ln u$, so we find:

$$\frac{r}{r_0} = \frac{f(1)}{f(u)}\exp(Ht), \qquad f(u) = \left[\eta + u + \sqrt{(\eta + 3u)(\eta - u)}\right]\exp\left(-\sqrt{\frac{\eta + 3u}{\eta - u}}\right) ;$$

$$u = r/r_0, \quad 0 < u \le 1 < \eta; \qquad \eta = 2^{2/3}r_0/r_\Lambda ; \qquad (4.34)$$

$$\phi - \phi_0 = \sqrt{3}\left(\sqrt{\frac{\eta + 3}{\eta - 1}} - \sqrt{\frac{\eta + 3u}{\eta - u}}\right) + \left(\arcsin\frac{\eta - 3}{2\eta} - \arcsin\frac{\eta - 3u}{2\eta}\right)$$

Of course, this escape solution is asymptotically radial and obeys the Hubble law (4.9). There are just two free initial parameters in it, r_0 and ϕ_0, of which only ϕ_0 is completely arbitrary (and can always be set to zero by counting the polar angle from this initial value). The starting radius is bounded from below: $\eta > 1$, therefore

$$r_0 > 2^{-2/3}r_\Lambda \equiv r_{min}. \qquad (4.35)$$

The trajectory cannot start too close to the center, but its beginning can be inside the equilibrium sphere $r = r_\Lambda$, in which case it crosses the sphere at a finite time with a non-zero velocity.

Another story unveils when the initial radial velocity is negative, $\dot{r}(0) < 0$, as are the signs in the solution (4.6). In this case u goes from 1 to η (r from r_0 to $r_{min} = r_0/\eta$). Since both above integrals diverge in the limit $u \to \eta - 0$, $r(t)$ never actually reaches r_{min}, while the phase $\phi(t)$ grows infinitely. This means that the orbit spirals all the way down to r_{min}, as shown in Fig. 4.7.

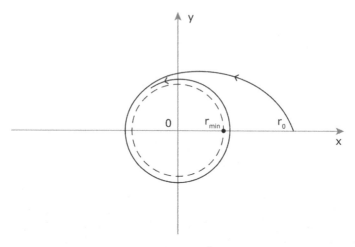

Fig. 4.7 Spiral orbit for $E = E^*$, $L = L_*$, and $E = E''$, $0 < L < L_*$ ($r_0 > r_{min}$, $\phi_0 = 0$)

The solution can be written by just changing t to $-t$ in the formula (4.34). But now the leading term in the limit $u \to \eta - 0$ is $(\eta - u)^{-1/2}$, which relates to a more adequate representation, better depicting the motion:

$$\frac{r}{r_{min}} = \frac{[Ht + g(u)]^2 + 1}{[Ht + g(u)]^2 - 1}, \quad g(u) = \sqrt{\frac{\eta + 3}{\eta - 1}} + \ln \frac{\eta + u + \sqrt{(\eta + 3u)(\eta - u)}}{u\left[\eta + 1 + \sqrt{(\eta + 3)(\eta - 1)}\right]} ;$$

$$u = r/r_0, \quad 1 \leq u < \eta; \quad \eta = 2^{2/3} r_0/r_\Lambda, \quad r_{min} = r_0/\eta = 2^{-2/3} r_\Lambda ; \qquad (4.36)$$

$$\phi - \phi_0 = \sqrt{3}\left(\sqrt{\frac{\eta + 3u}{\eta - u}} - \sqrt{\frac{\eta + 3}{\eta - 1}}\right) + \left(\arcsin \frac{\eta - 3u}{2\eta} - \arcsin \frac{\eta - 3}{2\eta}\right) .$$

The corresponding bounded orbit with the special initial parameters (4.33) can again start from any $r_0 > 2^{-2/3} r_\Lambda$, but, instead of escaping to infinity, it spirals down to the radius r_{min}: as seen from the formula (4.36), $r(t)$ is always larger than r_{min} and tends to it at large times. The validity of the initial condition is not so obvious, but taking into account

$$g^2(1) = \frac{\eta + 3}{\eta - 1} ,$$

one sees that $r(0) = r_0$ does hold. A useful asymptotic approximation

$$r(t) = r_{min}\left[1 + 2/(Ht)^2 + \cdots\right]$$

works well for large times, $Ht \gg 1$. It can be easily extended to a series solution for $r(t)$, like in Sect. 3.4.1; then one can also get a series solution for $\phi(t)$.

4.5.3 The Case $E = E'$, $0 < L < L_*$

It is characterized by the dimensionless parameters $a = a' = -\alpha'/l$, $0 < b < b_*$. As seen in Fig. 4.6a, the polynomial $Q(u)$ has a double root $u_1 = u_2$ which also coincides with u_+; so, by the formula (4.11),

$$u_{1,2} = u_+(a') = \gamma_+/l, \qquad \gamma_+ = \left(3 + \sqrt{9 - 32\alpha'}\right)/8 . \tag{4.37}$$

and

$$Q(u) = l(u + \xi)(\eta - u)(u - u_+)^2 .$$

It is important that α' and hence γ_+ depends only on $\beta = l^3 b$, and thus on L, but not on r_0, since $l^3 b \propto (L^2/GMr_\Lambda)^3$.

The identities (4.22) become

$$\eta + (-\xi) = 1/l - 2u_+, \qquad (-\xi)\eta = -b/lu_+^2 ,$$

giving the explicit expressions for ξ and η:

$$\xi, \eta = \frac{1}{2}\left[\sqrt{\left(\frac{1}{l} - 2u_+\right)^2 + \frac{4b}{lu_+^2}} \mp \left(\frac{1}{l} - 2u_+\right)\right] =$$
$$\frac{1}{2l}\left[\sqrt{(1 - 2\gamma_+)^2 + \frac{4\beta}{\gamma_+^2}} \mp (1 - 2\gamma_+)\right] \equiv \frac{\zeta_\mp(\beta)}{l} ; \tag{4.38}$$

here minus corresponds to ξ, plus to η, and $\gamma_+ = \gamma_+(\beta)$ is defined by the Eq. (4.37).

As in the previous section, the only interval allowed for motion is $0 < u < \eta$, but now there is just one infinite solution. The reason is that here η is a simple, not a double, root of $Q(u)$, and thus it is a turning point of the radial motion. When the initial radial velocity is positive, the particle goes straight to infinity, same as in the solution (4.34). However, in the opposite situation $\dot{r}(0) < 0$, the particle, moving towards the center, reaches the minimum radius $r_{min} = r_0/\eta$ in a finite time, bounces radially, and then escapes to infinity as in the first case. We find r_{min} by virtue of the expression (4.38) for η and the definition (4.8) of l:

$$r_{min} = \frac{r_0}{\eta} = \frac{lr_0}{\zeta_+} = \frac{L^2}{2GM\zeta_+} = \frac{3}{2^{8/3}\zeta_+}\left(\frac{L}{L_*}\right)^2 r_\Lambda . \tag{4.39}$$

As $r_0 \geq r_{min}$, the motion cannot start too close to the center, but it can begin inside the equilibrium sphere $r = r_\Lambda$.

The needed integrals

$$\sqrt{b} \int_u^1 \frac{du}{u\sqrt{Q(u)}} = \sqrt{\frac{b}{l}} \int_u^1 \frac{du}{u(u_+ - u)\sqrt{(u + \xi)(\eta - u)}} =$$

$$\frac{1}{u_+}\sqrt{\frac{b}{l}} \left[\int_u^1 \frac{du}{u\sqrt{(u + \xi)(\eta - u)}} + \int_u^1 \frac{du}{(u_+ - u)\sqrt{(u + \xi)(\eta - u)}} \right] ;$$

$$\sqrt{l} \int_u^1 \frac{u\,du}{\sqrt{Q(u)}} = \int_u^1 \frac{u\,du}{(u_+ - u)\sqrt{(u + \xi)(\eta - u)}} =$$

$$-\int_u^1 \frac{du}{\sqrt{(u + \xi)(\eta - u)}} + u_+ \int_u^1 \frac{du}{(u_+ - u)\sqrt{(u + \xi)(\eta - u)}}$$

are easily calculated, for instance, by the formulas **380.111, 380.001** from [31]. Taking into account the second of identities (4.22) written as $(1/u_+)\sqrt{b/l\xi\eta} = 1$, we obtain the following escape solution for $\dot{r}(0) > 0$:

$$\frac{r}{r_0} = \frac{f(1)}{f(u)} \exp(Ht), \qquad f(u) = \kappa(u) \exp \psi(u) ;$$

$$\phi - \phi_0 = [\arcsin \theta(1) - \arcsin \theta(u)] + [\psi(u) - \psi(1)] ;$$

$$u = r/r_0, \quad 0 < u \le 1 < \eta; \quad r_0 \ge r_{min} ; \tag{4.40}$$

$$\kappa(u) = 2\sqrt{\xi\eta(u + \xi)(u_+ - u)} + (\eta - \xi)u + 2\xi\eta; \qquad \theta(u) = \frac{\eta - \xi - 2u}{\xi + \eta} ;$$

$$\psi(u) = \sqrt{\frac{\xi\eta}{(u_+ + \xi)(u_+ - \eta)}} \arcsin \left[\frac{2u_+ + \xi - \eta}{\xi + \eta} - \frac{2(u_+ + \xi)(u_+ - \eta)}{(\xi + \eta)(u_+ - u)} \right] ;$$

the values of $u_+ = u_+(a')$, ξ and η are found by the expressions (4.37) and (4.38).

When $\dot{r}(0) < 0$, the solution is at first given by the same formula (4.40) where $\exp(Ht)$ is replaced with $\exp(-Ht)$, and $1 \le u < \eta$. The moving point comes to the minimum radius, i.e. u reaches η, at the moment

$$t_b = \frac{1}{H} \ln \frac{\eta f(1)}{f(\eta)} ,$$

and bounces from it radially. After that the solution (4.40), in which t is replaced with $(t - t_b)$, and r_0 with r_{min}, takes hold.

Recall that the solution is valid under the restrictions

$$E = E' = -2\alpha' \left(\frac{GM}{L} \right)^2 , \qquad 0 < L < L_* = \sqrt{\frac{3}{2^{8/3}}} GMr_\Lambda$$

on its physical initial parameters.

4.5.4 The Case $E = E''$, $0 < L < L_*$

Here $a = a'' = -\alpha''/l$, $0 < b < b_*$, the polynomial $Q(u)$ has all four roots real, with the double root $\eta = u_1$ that coincides also with u_- and forms thus a zero minimum point [see Fig. 4.6c) and the expression (4.11)]:

$$\eta = u_1 = u_-(a'') = \frac{\gamma_-}{l}, \qquad \gamma_- = \frac{3 - \sqrt{9 - 32\alpha''}}{8} = \frac{4\alpha''}{3 + \sqrt{9 - 32\alpha''}} \; ; \quad (4.41)$$

of course, since $\alpha'' = \alpha''(\beta)$, also $\gamma_- = \gamma_-(\beta)$. Consequently, the polynomial reads:

$$Q(u) = l(u + \xi)(u - u_-)^2(u_2 - u) \; ;$$

the identities (4.22) become

$$u_2 + (-\xi) = 1/l - 2u_-, \qquad (-\xi)u_2 = -b/lu_-^2 \; ,$$

leading to the formulas for ξ and u_2:

$$\xi, u_2 = \frac{1}{2}\left[\sqrt{\left(\frac{1}{l} - 2u_-\right)^2 + \frac{4b}{lu_-^2}} \mp \left(\frac{1}{l} - 2u_-\right)\right] =$$

$$\frac{1}{2l}\left[\sqrt{(1 - 2\gamma_-)^2 + \frac{4\beta}{\gamma_-^2}} \mp (1 - 2\gamma_-)\right] \equiv \frac{\chi_\mp}{l} \; , \qquad (4.42)$$

in a perfect similarity with expressions (4.38) of the previous section. So again, and as in all cases considered, the four roots of $Q(u)$ are found explicitly through the parameters a, b, and l.

This case, however, is rather rich, because two segments, $0 < u < u_-$ and $u_- < u < u_2$ are allowed for motion, which results in one infinite trajectory and two bounded spiral orbits.

We start with the first segment, $0 < u < u_-$. The allowed radii are bounded from below by the radius r_{min}, which, by the expression (4.41) and the definition (4.8), is:

$$r_{min} = \frac{r_0}{u_-} = \frac{lr_0}{\gamma_-} = \frac{L^2}{2GM\gamma_-} = \frac{3}{2^{8/3}\gamma_-}\left(\frac{L}{L_*}\right)^2 r_\wedge \; . \qquad (4.43)$$

Since, according to the first and second expressions (4.41), $\gamma_- < 3/8$ and $\gamma_- > 2\alpha''/3$, the following bounds for r_{min} hold:

$$2^{1/3}\left(\frac{L}{L_*}\right)^2 r_\wedge < r_{min} < \frac{9}{2^{11/3}\alpha''}\left(\frac{L}{L_*}\right)^2 r_\wedge \; . \qquad (4.44)$$

The upper bound of the minimum radius, by the formulas (4.17) and (4.19), simplifies to l.o. in the parameter β, and becomes independent of the orbital momentum:

$$r_{min} < \frac{3}{2^{11/3}\alpha''}\left(\frac{L}{L_*}\right)^2 r_\wedge \approx \frac{3}{8\beta^{1/3}}\left(\frac{L}{L_*}\right)^2 r_\wedge = 2\,r_\wedge\,. \tag{4.45}$$

The integrals (4.6) needed here are:

$$\sqrt{b}\int_u^1 \frac{du}{u\sqrt{Q(u)}} = \sqrt{\frac{b}{l}}\int_u^1 \frac{du}{u(u_- - u)\sqrt{(u+\xi)(u_2-u)}}\,;$$

$$\sqrt{l}\int_u^1 \frac{u\,du}{\sqrt{Q(u)}} = \int_u^1 \frac{u\,du}{(u_- - u)\sqrt{(u+\xi)(u_2-u)}}\,. \tag{4.46}$$

They appear to be exactly as in Sect. 4.5.3, with just u_- replacing u_+, and u_2 replacing η; however, the relation between the parameters is different, $u_+ > \eta$ in one case, and $u_- < u_2$ in the other. That makes the difference in the final expressions; as before, to calculate the first integral we use the identity (4.22), reading here $(1/u_-)\sqrt{b/l\xi u_2} = 1$.

The picture of motion, is, in fact, like that in Sect. 4.5.2: for $\dot{r}(0) > 0$ the particle escapes to infinity, while for the opposite case $\dot{r}(0) < 0$ it spirals down to the circle $r = r_{min}$.

The result for the infinite trajectory is:

$$\frac{r}{r_0} = \frac{f(1)}{f(u)}\exp{(Ht)}, \qquad f(u) = \kappa(u)\left[\psi(u)(u_- - u)\right]^\sigma\,;$$

$$\phi - \phi_0 = \arcsin\theta(1) - \arcsin\theta(u) + \frac{u_-}{\sqrt{(\xi + u_2)(u_2 - u_-)}}\ln\frac{\psi(1)(u_- - u)}{\psi(u)(u_- - 1)}\,;$$

$$u = r/r_0,\quad r_0 > r_{min},\quad 0 < u \le 1 < u_- = \eta;\qquad \sigma = \sqrt{\frac{\xi u_2}{(\xi + u_2)(u_2 - u_-)}}\,; \tag{4.47}$$

$$\kappa(u) = 2\sqrt{\xi u_2(u+\xi)(u_2-u)} + (u_2 - \xi)u + 2\xi u_2;\qquad \theta(u) = \frac{u_2 - \xi - 2u}{\xi + u_2}\,;$$

$$\psi(u) = 2\sqrt{p(u+\xi)(u_2-u)} + q(u_- - u) + 2p\,;$$

$$p = (\xi + u_2)(u_2 - u_-),\ q = 2u_- + \xi - u_2\,.$$

Here $u_- = u_-(a'')$ is given by the formula (4.41), and ξ and u_2 are specified by the expressions (4.42). The initial radius r_0 must be larger than r_{min} determined by Eq. (4.44). This form of solution is based on the large time leading term $\ln u$ in the first of the above integrals.

If, contrary to the above, the initial radial velocity is negative, $\dot{r}(0) < 0$, then u tends from below to u_- at large times, so both integrals diverge in this limit. The proper form of the solution based on the large time leading term $\ln(u_- - u)$ reads:

$$\frac{r}{r_{min}} = \left\{ 1 - \frac{u_- - 1}{u_-} \left[\frac{\psi(1)}{\psi(u)} \right] \left[\frac{u\kappa(1)}{\kappa(u)} \right]^{1/\sigma} \exp\left(-Ht/\sigma\right) \right\}^{-1} ;$$

$$\phi - \phi_0 = \frac{u_-}{\sqrt{(\xi + u_2)(u_2 - u_-)}} \ln \frac{\psi(u)(u_- - 1)}{\psi(1)(u_- - u)} + \arcsin \theta(u) - \arcsin \theta(1) ;$$

$$u = r/r_0, \quad r_0 > r_{min}, \quad 1 \le u < u_- = \eta , \tag{4.48}$$

with the same values of $\psi(u)$, $\kappa(u)$, $\theta(u)$ and σ as in the solution (4.47). Formulas (4.48) describe a spiral orbit of the same kind as in Sect. 4.5.2 shown in Fig. 4.7. The only significant difference between the two is a much faster approach to the limit circle $r = r_{min}$: here the distance between the circle and the moving point goes down exponentially, $(r - r_{min}) \propto \exp\left(-Ht/\sigma\right)$, instead of the inverse square, $(r - r_{min}) \propto (Ht)^{-2}$, as in the solution (4.36).

Finally, we go for the second segment allowed for motion,

$$u_- < u \le u_2, \quad \text{or} \quad r^{min} \le r < r^{max}, \quad r^{min} = r_0/u_2, \quad r^{max} = r_0/u_- ;$$

the maximum radius here coincides, naturally, with the minimum radius of the first segment, $r^{max} = r_{min}$, given by the equality (4.44). The minimum radius is found according to the formula (4.42),

$$r^{min} = \frac{r_0}{u_2} = \frac{lr_0}{\chi_+} = \frac{L^2}{2GM\chi_+} = \frac{3}{2^{8/3}\chi_+} \left(\frac{L}{L_*} \right)^2 r_\Lambda, \tag{4.49}$$

$$\chi_+(\beta) = \frac{1}{2} \left[\sqrt{(1 - 2\gamma_-)^2 + \frac{4\beta}{\gamma_-^2}} + (1 - 2\gamma_-) \right] ,$$

with $\gamma_-(\beta)$ defined by Eq. (4.41). Apparently, the solution represents a bounded orbit spiraling up to r^{max}. Its representation is almost the same as the solution (4.48), with just a slightly different way of writing it:

$$\frac{r}{r^{max}} = \left\{ 1 + \frac{1 - u_-}{u_-} \left[\frac{\psi(1)}{\psi(u)} \right] \left[\frac{u\kappa(1)}{\kappa(u)} \right]^{1/\sigma} \exp\left(-Ht/\sigma\right) \right\}^{-1} ; \tag{4.50}$$

$$\phi - \phi_0 = \frac{u_-}{\sqrt{(\xi + u_2)(u_2 - u_-)}} \ln \frac{\psi(u)(u_- - 1)}{\psi(1)(u_- - u)} + \arcsin \theta(u) - \arcsin \theta(1) ;$$

$$u = r/r_0, \quad \eta = u_- < u \le 1 < u_2; \quad r^{min} \le r_0 < r^{max}, \quad \dot{r}(0) > 0 .$$

All quantities involved here are defined above. In the case of a negative initial radial velocity, $\dot{r}(0) < 0$, the point first moves towards r^{min}, reaches it in a final time, bounces radially, and then follows the trajectory (4.50). Spiral orbits for both signs of the initial radial velocity are plotted in Fig. 4.8.

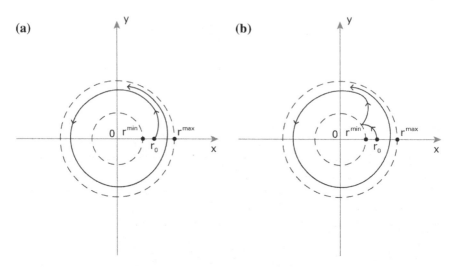

Fig. 4.8 Finite spiral orbits for $E = E''$, $0 < L < L_*$ ($r^{min} \leq r_0 < r^{max}$, $\phi_0 = 0$) **a** $\dot{r}(0) > 0$; **b** $\dot{r}(0) < 0$

4.6 Orbital Solutions Expressed Through Elliptic Integrals

We start with generic finite motions existing for $E' < E < E'' < 0$, $0 < L < L_*$. In terms of dimensionless parameters, the range is $a' < a < a'' < 0$, $0 < b < b_*$, so the polynomial $Q(u)$ has four real roots, $-\xi < 0 < \eta < u_1 < u_2$, Fig. 4.6b. Its proper representation for the interval $u_1 \leq u \leq u_2$ ($r_{min} \leq r \leq r_{max}$, with $r_{min} = r_0/u_2$, $r_{max} = r_0/u_1$), where all generic bounded orbits reside, is

$$Q(u) = l(u + \xi)(u - \eta)(u - u_1)(u_2 - u) .$$

The integrals (4.6) representing the solution can be reduced to standard elliptic integrals with the help of Table 1 from [32], Sect. 13.5. The result is expressed through the elliptic integral (3.15) of the third kind, and the elliptic integral $F(x, k)$ of the first kind defined as (see [31], **770**)

$$F(x, k) = \int\limits_{0}^{x} \frac{dx}{\sqrt{(1 - x^2)(1 - k^2 x^2)}} .$$

The result reads:

$$Ht = \pm \frac{2}{\eta} \sqrt{\frac{b}{l(u_2 - \eta)(u_1 + \xi)}} \times$$

$$\left\{ [F(X(1), k) - F(X(u), k)] - \frac{u_1 - \eta}{u_1} [\Pi(X(1), \nu_1, k) - \Pi(X(u), \nu_1, k)] \right\} ;$$

$$\phi - \phi_0 = \pm \frac{2}{\sqrt{(u_2 - \eta)(u_1 + \xi)}} \times \qquad (4.51)$$

$$\{ \eta [F(X(1), k) - F(X(u), k)] - (u_1 - \eta) [\Pi(X(1), \nu_2, k) - \Pi(X(u), \nu_2, k)] \} ;$$

$$u = r/r_0, \quad u_1 \le u \le u_2; \quad r_{min} \le r \le r_{max}; \quad X(u) = \sqrt{\frac{u_2 - \eta}{u_2 - u_1} \frac{u - u_1}{u - \eta}} ;$$

$$k^2 = \frac{\eta + \xi}{u_1 + \xi} \frac{u_2 - u_1}{u_2 - \eta} < 1; \quad \nu_1 = -\frac{\eta}{u_1} \frac{u_2 - u_1}{u_2 - \eta} < -1; \quad \nu_1 = -\frac{u_2 - u_1}{u_2 - \eta} < -1 .$$

As usual, the plus sign corresponds to the motion from r_{min} to r_{max}, the minus to the motion in the opposite direction. Note that $X(u)$ grows monotonically from $X(u_1) = 0$ to $X(u_2) = 1$, so $0 < X(u) < 1$ for $u_1 < u < u_2$, and the elliptic integrals are well defined on this interval. The expressions for $u_{1,2}$ through ξ and η are given by the formula (4.23). More effective, though, are the asymptotic expressions (B.6) derived in Appendix B ($\alpha \ne \frac{1}{4}$):

$$u_1 = \frac{1}{2l} \left[\left(1 - \sqrt{1 - 4\alpha}\right) - \frac{2}{\sqrt{1 - 4\alpha}\left(1 - 2\alpha - \sqrt{1 - 4\alpha}\right)} \beta + \cdots \right] ;$$

$$u_2 = \frac{1}{2l} \left[\left(1 + \sqrt{1 - 4\alpha}\right) + \frac{2}{\sqrt{1 - 4\alpha}\left(1 - 2\alpha + \sqrt{1 - 4\alpha}\right)} \beta + \cdots \right] ;$$

they allow us to estimate the radii r_{min} and r_{max} of the annulus within which the orbit lies. We have, to lowest order:

$$r_{min} = \frac{r_0}{u_2} \approx \frac{2lr_0}{1 + \sqrt{1 - 4\alpha}} = \frac{3}{2^{8/3}} \left(\frac{L}{L_*}\right)^2 \frac{r_\Lambda}{1 + \sqrt{1 - 4\alpha}} ,$$

where we used the definition (4.8) of l and (4.18) of L_*. Thus the minimum radius, proportional to $(L/L_*)^2$, is within

$$\frac{3}{2^{11/3}} \left(\frac{L}{L_*}\right)^2 r_\Lambda < r_{min} < \frac{3}{2^{8/3}} \left(\frac{L}{L_*}\right)^2 r_\Lambda . \qquad (4.52)$$

In the same fashion, for the maximum radius we derive, to l.o.:

$$r_{max} = \frac{r_0}{u_1} \approx \frac{3}{2^{8/3}} \left(\frac{L}{L_*}\right)^2 \frac{r_\Lambda}{1 - \sqrt{1 - 4\alpha}} =$$

$$\frac{3}{2^{8/3}} \left(\frac{L}{L_*}\right)^2 \frac{1 + \sqrt{1 - 4\alpha}}{4\alpha} r_\Lambda < \frac{3}{2^{11/3}} \left(\frac{L}{L_*}\right)^2 \frac{r_\Lambda}{\alpha} \ .$$

As $\alpha > \alpha'' > 0$, using the l.o. expression (4.17) for $\alpha''(\beta)$ and formula (4.19) for β we arrive at the desired estimate:

$$r_{max} < \frac{3}{2^{11/3}} \left(\frac{L}{L_*}\right)^2 \frac{r_\Lambda}{\alpha''} \lesssim \frac{3}{2^{11/3}} \left(\frac{L}{L_*}\right)^2 \frac{2^{2/3} r_\Lambda}{3\beta^{1/3}} = \frac{2}{3} r_\Lambda \ . \qquad (4.53)$$

This significant estimate demonstrates that all generic bounded orbits are located within $r = 2r_\Lambda/3$, well inside the no-gravity sphere.

The solution (4.51) is periodic under the condition (4.25), which in this case reads:

$$\frac{\pi m}{n} = \frac{2}{\sqrt{(u_2 - \eta)(u_1 + \xi)}} \times \qquad (4.54)$$

$$\{\eta \left[F(X(u_2), k) - F(X(u_1), k)\right] - (u_1 - \eta) \left[\Pi(X(u_2), \nu_2, k) - \Pi(X(u_1), \nu_2, k)\right]\} =$$

$$\frac{2}{\sqrt{(u_2 - \eta)(u_1 + \xi)}} \left[\eta K(k) + (u_1 - \eta)\Pi(1, \nu_2, k)\right] \ .$$

Here $m, n > 0$ are integers, we used $X(u_1) = 0$, $X(u_2) = 1$, and the notation

$$K(k) = \int\limits_0^1 \frac{dx}{\sqrt{(1 - x^2)(1 - k^2 x^2)}}$$

stands for the complete elliptic integral of the first kind ([31], **773.1**). Since all the roots of $Q(u)$, and thus all the parameters involved above, are functions of the polynomial coefficients l, a and b, the above condition can be written symbolically,

$$\mathcal{F}(a, b, l) = \frac{\pi m}{n} \ , \qquad (4.55)$$

with some function $\mathcal{F}$. Any triplet $\{a, b, l\}$ satisfying this strongly nonlinear equation for given m/n corresponds to a closed orbit, provided that it obeys the conditions $l > 0$, $0 < b < b_*$, $a' < a < a''$, or

$$l > 0, \quad 0 < l^3 b < (3/16)^3, \quad a'(b, l) < a < a''(b, l) \ .$$

At most, one obtains thus a countable set of 2D manifolds (surfaces) $l = l_{mn}(a, b)$ in the 3D space of initial data parameterized by a, b and l. Any proper triplet, i.e., any point on one of these surfaces, gives rise to a periodic solution. Its period is given by the formula (4.26), namely,

$$T_{mn} = \frac{4n}{H\eta} \sqrt{\frac{b}{l(u_2 - \eta)(u_1 + \xi)}} \left[\eta K(k) - \frac{u_1 - \eta}{u_1} \Pi(1, \nu_1, k) \right] ;$$

the dependence on m is hidden in the values of parameters.

Similar expressions of the solutions describing generic finite motions were obtained in paper [14], however, without indication of the energy and angular momentum ranges within which they exist. Non-generic infinite motions that exist in the same ranges were also given there in terms of elliptic functions.

Infinite motions, which exist for all energies and angular momenta, can be expressed through elliptic integrals in a similar fashion. The generic case is when there are no other motions, i.e., the polynomial $Q(u)$ has just two real roots, $-\xi$ and η, while $u_{1,2}$ are complex conjugate; this excludes just the range $E' \leq E \leq E''$, $0 < L < L_*$. In this case the integrals (4.6) are again reduced to a combination of elliptic integrals of the 1st and 3rd kind, as well as logarithmic functions, with the help of Table 2 from [32], Sect. 13.5. However, the result is rather cumbersome, so we choose not to give it here.

Chapter 5
All Motions: Summary. Locality and Stability of Finite Motions

Abstract We summarize the properties of all found solutions, and show that finite motions are essentially localized, predominantly within the no-gravity sphere. We then discuss their stability and show that, with the exception of generic periodic and aperiodic motions described by elliptic integrals, finite motions are structurally unstable.

5.1 All Motions

Here we summarize the general results of our investigation of all possible motions. There are five groups of finite motions listed below.

1. Radial ($L = 0$) fall on the center, possible for the whole range of total energy, $-\infty < E < \infty$. For $E > U_{max} = U(r_\Lambda)$ it requires the negative initial radial velocity, $\dot{r}(0) < 0$; when $E = U_{max}$, also the initial position must be bounded, $\dot{r}(0) < 0$, $r_0 < r_\Lambda$; for $E < U_{max} < 0$ only the initial position is restricted, $r_0 < r_-(E)$ (see Fig. 2.1).

2. Radial ($L = 0$) approach to an equilibrium at $r = r_\Lambda$ possible for $E = U_{max}$, with $\dot{r}(0) > 0$ when $r_0 < r_\Lambda$, and $\dot{r}(0) < 0$ when $r_0 > r_\Lambda$.

3. Spiral orbits possible for

    (a) $E = E_* < 0$, $L = L_*$, with $r_0 > 2^{-2/3}r_\Lambda$, $\dot{r}(0) < 0$;

    (b) $E = E'' < 0$, $0 < L < L_*$ (r_0, $\dot{r}(0)$ are restricted in Sect. 4.5.4).

4. Circular orbits possible for $0 < L \le L_*$, $E \le -3(GMH/2)^{2/3}$ inside the equilibrium sphere, $r < r_\Lambda$.

5. Orbits radially oscillating, in a periodic or aperiodic way, between two circles. Possible for $E' < E < E'' < 0$, $0 < L < L_*$. Periodic solutions exist for at most countable set of surfaces in the space of initial parameters a, b, L, see Sect. 4.6.

All other solutions represent infinite motions, either radial or asymptotically radial; they exist for any energy and angular momentum, $|E| < \infty$, $0 \le L < \infty$. Figure 5.1 shows the distribution of finite and infinite motions in the E, L plane of initial data. All finite motions are in the shaded domain, infinite motions are possible in the whole upper half-plane. Circular orbits are in the semi-strip $E < -3GM/2^{2/3}r_\Lambda$,

A. Silbergleit and A. Chernin, *Kepler Problem in the Presence of Dark Energy, and the Cosmic Local Flow*, SpringerBriefs in Physics, https://doi.org/10.1007/978-3-030-36752-7_5

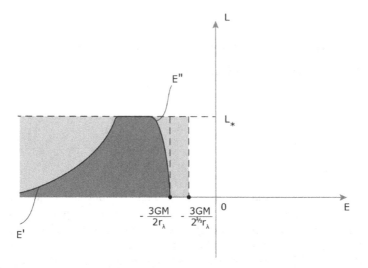

Fig. 5.1 Picture of orbital motions in the E, L plane

other finite motions are in the dark-shaded region; initial radius is properly restricted in every case.

5.2 Localization of Finite Motions

We see that finite motions, or bounded orbits, exist for negative total energies, $E < 0$, and finite orbital momenta, $0 \leq L \leq L_*$, with the only exception of the fall on the center with $E > 0$. The radial fall trajectories with positive or negative energies satisfying $E > U_{max}$ can start at any $r_0 > 0$, but even in such case the particle gets inside the equilibrium sphere $r = r_\Lambda$ in a finite time. Otherwise all our findings show that finite motions, i.e., bounded orbits, are essentially localized in space, predominantly, inside this no-gravity sphere.

Indeed, the radial fall on the center with $E < U_{max} < 0$ occurs for $r < r_\Lambda$. The spiral orbit of Sect. 4.2 converges on the circle of the radius $(r_\Lambda/2^{2/3}) < r_\Lambda$ from the outside, so at most its final part lies outside the equilibrium sphere, if the particle starts at $r_0 > r_\Lambda$; it stays there for a finite time only. All circular orbits examined in Sect. 4.1 have their radii below r_Λ. A particle on a generic finite orbit oscillates in the radial direction between two circles of the radius $r_{min} < r_{max}$. According to the estimate (4.53), $r_{max} < 2/3 r_\Lambda$, so the orbits reside well inside the equilibrium sphere.

Hence the only exception is the two spiral orbits of Sect. 4.4, which both converge on the same circumference $r = r_{min}$, one from the outside, the other from the inside. This radius is estimated in the formula (4.45), its upper bound independent of the initial parameters is $r_{min} < 2r_\Lambda$. In any case, a particle on either orbit spends an infinite time close to r_{min}, inside the sphere of the radius $2r_\Lambda$, so these orbits are still well localized. Some of the results on the localization of finite motions are found in paper [14].

5.3 Stability of Finite Motions

Our final point here is that almost all finite solutions, except generic orbits of group 5 above, are structurally unstable, that is, turn to infinite solutions under small (infinitesimal) perturbations. Clearly, the solutions of groups 1–3 exist only for special values of energy, or angular momentum, or both, so they turn to infinite motions with any change of these initial parameters.

Instability of circular orbits (group 4) was mentioned in paper [14] and analyzed recently in [20]. Here we carry out this analysis in more explicit and detailed way. We start with disturbances in the plane of motion $z = 0$. Azimuthal perturbations do not change the shape of the orbit, so it is fortunate that we can study radial perturbations separately. Namely, by differentiating the energy conservation equation (4.2) in time we obtain the radial motion equation:

$$\ddot{r} - \frac{L^2}{r^3} + \frac{GM}{r^2} - H^2 r = 0 \; .$$

Setting $r(t) = r_0 + \delta r(t)$, where $r_0 = $ const is the orbit radius, and $\delta r(t)$ is a small perturbation, to linear order in it we find [see Eq. (4.29)]:

$$\ddot{\delta r} + \left(\frac{3L^2}{r_0^4} - \frac{2GM}{r_0^3} - H^2 \right) \delta r = 0, \quad \text{or} \quad \ddot{\delta r} + \left(\frac{GM}{r_0^3} - 4H^2 \right) \delta r = 0. \quad (5.1)$$

The coefficient in this equation is negative, zero, or positive if, respectively, r_0 is larger, equal to, or smaller than $4^{-1/3} r_\Lambda$; remarkably, this is the radius of the circular orbit with the maximum angular momentum $L = L_*$, as shown in Sect. 4.5.

Therefore the circular orbit of the radius r_0 is unstable to in-plane perturbations for $4^{-1/3} r_\Lambda \le r_0 < r_\Lambda$, and neutrally stable if $0 < r_0 < 4^{-1/3} r_\Lambda$ (small oscillations about the unperturbed trajectory). The instability is exponential in time for $4^{-1/3} r_\Lambda < r_0 < r_\Lambda$, and linear in time for $r_0 = 4^{-1/3} r_\Lambda$. The characteristic time of the exponential instability, according to the Eq. (5.1) and expression (4.27), is

$$\tau_{ins} = \left(4H^2 - \frac{GM}{r_0^3} \right)^{-1/2} = \left(3H^2 - \omega^2 \right)^{-1/2} \; .$$

So it tends to $(\sqrt{3}H)^{-1}$ when the unperturbed orbit goes close to the no-gravity sphere, $r_0 \to r_\Lambda - 0$, and the orbital frequency ω tends to zero. Under the real conditions met in our universe, $\tau_{ins} \lesssim (\sqrt{3}H)^{-1}$ is huge in this case, so the perturbed circular orbits in a thin layer immediately inside the no-gravity sphere spend astronomically large times there, moving slowly on very tight and slowly unwinding spiral trajectories. Given the realistic observation times, those objects will not exhibit any detectable motion at all, just 'staying in place'.

To complete the stability analysis one should still consider the out-of-plane perturbations. As one expects, circular orbits prove to be neutrally stable to them, because the z-projection of the motion Eq. (2.6) to l.o. in the small perturbation $\delta z(t)$ reduces to

$$\ddot{\delta z} + \omega^2 \delta z = 0 ,$$

demonstrating small oscillations about the motion plane with the frequency ω. Our analysis shows thus that circular orbits with $4^{-1/3}r_\Lambda \leq r_0 < r_\Lambda$ are structurally unstable, while being structurally stable in the opposite case $0 < r_0 < 4^{-1/3}r_\Lambda$; this result appears to be new.

Finally, generic finite motions of Sect. 4.6 are structurally stable because it takes a *finite* variation in E and/or L to destroy them, turning into infinite motions.

Chapter 6
Dark Energy and the Local Cosmic Flow

Abstract We survey the work on the role of dark energy in local galactic flows based on recent observations and theoretical results. We start with the Local Group of galaxies (up to 3 Mpc), and then go to the neighboring local groups (10–30 Mpc). In all cases position-velocity data outside the corresponding zero-gravity spheres exhibit strong agreement with the Hubble law valid due to the DE presence. The DE density thus measured comes close to its cosmological value. Next we discuss the relative role of matter (gravity) and DE (anti-gravity) in the elements of the Cosmic Web. DE somewhat dominates 3D clusters of galaxies, and strongly dominates 2D superclusters (pancakes); alternatively, matter strongly prevails DE in large practically 1-dimensional filaments.

In this chapter we review recent results on local cosmology related to the two-body problem with dark energy, which was treated in the previous chapters. The review is mainly based on the papers mentioned in the Introduction, particularly, on [8, 10, 11, 13, 15–17], and on two talks that one of the authors (A. Ch.) gave at the Ginzburg Centennial Conference on Physics (May–June 2017, Moscow, Russia), and 6th Gamow International Conference (August 2019, Odessa, Ukraine).

6.1 Local Group of Galaxies

Local Group (LG) consists of two giant galaxies, our Milky Way, and the Andromeda galaxy (M31), now 0.77 Mpc apart. The estimates of its mass vary, $M = (2 - 3) \times 10^{12} M_\odot$ (in solar masses), currently. The upper bound was higher a few years ago, when the motion of M31 relative to our galaxy was computed in [16] from the Kepler problem with and without DE present. In the former case the cosmological value of the DE density was used in calculations,

$$\rho_\Lambda \simeq 0.7 \times 10^{-29} \, \text{g/cm}^3 \, . \tag{6.1}$$

For $M = 3 \times 10^{12} M_\odot$ the orbit is shown in Fig. 6.1a. The classical Kepler ellipse is in thin line, while a solid curve is the trajectory taking into account DE. The

© The Author(s), under exclusive license to Springer Nature Switzerland AG 2019
A. Silbergleit and A. Chernin, *Kepler Problem in the Presence of Dark Energy, and the Cosmic Local Flow*, SpringerBriefs in Physics,
https://doi.org/10.1007/978-3-030-36752-7_6

(a)

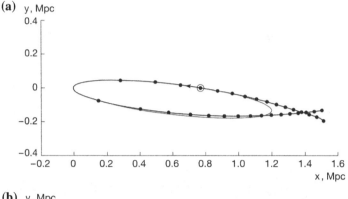

(b)

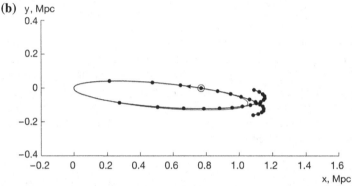

Fig. 6.1 Motion of the Andromeda galaxy relative to our galaxy. Thin line—classical elliptical orbit, solid line—orbit with DE taken into account. Points show successive positions with the interval of one billion years. The initial position is circled, the arrow shows the direction of motion. Total LG mass: **a** $3 \times 10^{12} M_\odot$; **b** $4 \times 10^{12} M_\odot$

difference between the two is significant in the distant past and future. Figure 6.1b gives the corresponding picture with $M = 4 \times 10^{12} M_\odot$, and it is surprising: not only the difference between no-DE and DE orbits is clearly seen at early and late times, but the system occurs to be unbound! The Andromeda galaxy is initially untied to our galaxy, then gets close to the classical Kepler orbit, stay in this quasi-bound state for billions of years, but eventually departs again. The transition value of the LG mass from the bound to unbound state is thus between $3 \times 10^{12} M_\odot$ and $4 \times 10^{12} M_\odot$; the very possibility of such transition is entirely due to the presence of dark energy in the volume occupied by the system.

For the following discussion we choose a moderate number for the LG mass,

$$M = 2 \times 10^{12} M_\odot .$$

Assuming the value (6.1) for the DE density we find the radius of the no-gravity sphere by its definition (2.7):

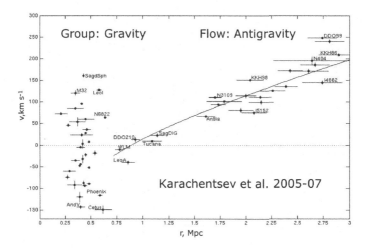

Fig. 6.2 Distance-velocity data for the local group of galaxies

$$r_\Lambda = 1.3\,\text{Mpc} \ .$$

The distance-velocity data for many galaxies in our and neighboring local groups were obtained by I. D. Karachentsev and his team in many observational sessions using Hubble Space Telescope (HST) and Large Altazimuth Telescope (BTA, a Latin transliteration of the Cyrillics acronym) during 2005–2017.

The LG data with error bars are plotted in Fig. 6.2. It shows that the pattern of data points distribution is different inside the zero-gravity sphere ($r < r_\Lambda$) and outside it. Inside, the spread of points along the velocity axis is large, they occupy the whole range between the minimum and maximum observed velocity more or less uniformly. It is a very different picture outside: data points appear to arrange themselves along some line. This becomes perfectly clear in the same, but clarified of redundant details, diagram of Fig. 6.3: for $r > r_\Lambda$ data points condense around nothing else than the straight line of the Hubble law, $v = Hr$. This is a bright manifestation of DE dominating the galactic dynamics: it provides the local flow with the regularity established by the Hubble law.

In Fig. 6.3 theoretical trajectories computed by problem (2.6) are also shown. Each data point lies on one of these trajectories (thin lines), which converge to the Hubble asymptote rather rapidly. The rate of convergence can be characterized by the relative width of the bunch as a function of distance,

$$\Delta(r) = \frac{v_{max}(r) - v_{min}(r)}{Hr} \ .$$

For example, at the beginning, when $r = 1.4\,\text{Mpc}$, we have $\Delta = 2.4$, while at just about twice the distance, $r = 3\,\text{Mpc}$, it drops more than 3 times to $\Delta = 0.74$.

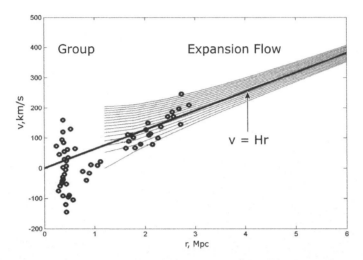

Fig. 6.3 Comparison of observational data and theoretical predictions for the local group. A bunch of trajectories outside the zero-gravity sphere is shown in thin lines. They converge to the straight line of the Hubble law $v = Hr$

The discussed observations imply another important conclusion. If dark energy does not interact with other types of matter, than it is perfectly uniform in the whole spacetime, its density is the same always and everywhere. This uniformity is usually assumed, as in the ΛCDM cosmological model (Λ is really a *constant*!). Here we also made such assumption by taking the local value of the DE density equal to the cosmological value (Eq. (6.1)). However, there is an alternative to this assertion: no reason is known that forbids DE-matter interaction, whose cosmological implications we discussed thoroughly in our book [4]. The above data show that the cosmological value of the Hubble parameter H works well for the local flow. They confirm the DE homogeneity providing the local value of DE density (at $\lesssim 3$ Mpc) very close to the cosmological one found at ~ 1000 Mpc.

6.2 Neighbor Local Groups of Galaxies

Moving up in the distance scale to 10–30 Mpc, we discuss the dynamics of neighboring local groups. A self-similar form the Hubble law,

$$\frac{v}{Hr_\Lambda} = \frac{r}{r_\Lambda} \, , \tag{6.2}$$

is instrumental for studying it. Of course, here r is the distance from the center of the respective group to the moving object. Since the masses of the groups are different, their no-gravity radii differ also; hence its own specific radius r_Λ should be used for

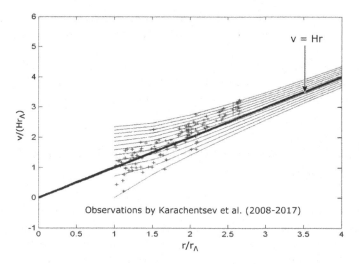

Fig. 6.4 Composite distance-velocity diagram with data of 125 galaxies in 4 local groups

each group in the formula (6.2). Contrary to that, the Hubble parameter value is the same for all of them if, as usually asserted, the DE density does not change locally. Otherwise, if DE interacts with matter, then the value of H generally changes from one group to another.

Four neighboring local groups in two clusters, *Virgo* and *Fornax*, were monitored. It took 330 orbits of HST during which the distances to approximately 300 galaxies within 11 Mpc from us were measured. Distance-velocity data for total of 125 galaxies in those four local groups are combined in a synthetic diagram of Fig. 6.4 plotted in dimensionless coordinates $\{r/r_\Lambda, v/Hr_\Lambda\}$. All parameters are calculated as explained above, and the universal value (6.1) of the DE density, hence of H, is taken. The alignment of data points with the Hubble straight line (6.2) is evident. It again supports the assumption of the DE uniformity, i.e., the proximity of locally determined values of ρ_Λ and H to their cosmological counterparts.

In connection with the discussed observations one theoretical result is worth mentioning. It is a generalization of the well-known cosmological random-motion theorem [36–38] to the case when dark energy is present, which was found in paper [39]. One of its implications is a practically important expression for the virial mass with the DE contribution (formula (27) in [39]).

It seems a good place for several conclusions drawn from the above discussion.
1. Anti-gravity caused by Einstein's cosmological constant is universal: it acts and may dominate dynamically on both global and local spatial scales, as predicted by General Relativity and discovered in observations with HST.
2. Local observations provide nearly the same value of the dark energy density, or Λ, as the large-scale cosmological determinations do.

3. A new phenomenon in the world of galaxies is predicted and discovered: nearly-linear self-similar local expanding flows of galaxies accelerated by Einstein's anti-gravity.

4. A new application of General Relativity is found, namely, to the world of galaxies around us, at the distances from 1 up to 100 Mpc. Observational confirmation of GR predictions provide yet another significant test of the theory validity.

6.3 Dark Energy Versus Matter in Various Structures of the Universe

Our previous considerations lead to an interesting general question: how do dark energy and matter combine in various large-scale formations existing in the universe, which of them prevails, in one situation or the other? One can quantify this interplay using the ratio of the DE density to matter density averaged over the structure volume,

$$ X = \frac{2\rho_\Lambda}{\langle \rho_{DM} \rangle + \langle \rho_{BM} \rangle + \langle \rho_R \rangle} = \frac{2\rho_\Lambda}{\langle \rho_M \rangle} \, , \tag{6.3} $$

where ρ_{DM} is the density of dark matter, ρ_{BM} is the density of baryonic luminous matter, ρ_R is the radiation density, and ρ_M is their total. The factor two is included in the numerator because of the effective DE density (2.2) producing antigravity. DE density ρ_Λ is assumed constant always and everywhere with the value (6.1). Matter density may change with the time, so also $X = X(t)$.

First of all, let us find out what is the value of X for the universe as a whole. According to the universally recognized ΛCDM model, currently DE comprises about 70% of the universe content, approximately 26% go for DM, and nearly all of the remaining 4% is baryonic matter, since only something like 0.01% falls to radiation. (The last number shows that radiation contribution should be neglected always except early universe, when it dominates everything else.) Due to these numbers, for the universe we find that now

$$ X_{univ} \approx 4.7 \, . $$

Since matter density drops with the time, the ratio grows in the future and decreases towards the past; the ΛCDM model gives $X = 1$ at the cosmic age of approximately 7 Gyr, about half of the current lifetime of the universe.

Next we discuss the building blocks of the Cosmic Web, which was first described a decade ago (see [35] and the references therein) as a vast network formed by all the galaxies of the universe. There are three types of structures in the Web: (1) groups and clusters of galaxies, of 1–10 Mpc size; (2) superclusters, or pancakes, 10–20 Mpc; (3) filaments with the length of 10–30 Mpc (we studied formation of all three elements based on exact Newtonian collapse solutions in [40]). The blocks differ geometrically: clusters are essentially 3-dimensional, superclusters are practically 2-dimensional, and filaments are nearly 1-dimensional, as implied by their name.

This geometrical characterization prompts suggesting that they also differ by their DE/matter content: intuitively, $X \sim 1$ for clusters, $X \gg 1$ for pancakes, and $X \ll 1$ for filaments. The following three examples support this conjecture.

6.3.1 Local Groups (3D)

A good model for our and other local groups is a ball of the zero-gravity radius r_Λ, whose volume, by the definition (2.7), is

$$V_\Lambda = \frac{4\pi}{3} r_\Lambda^3 = \frac{M}{2\rho_\Lambda} .$$

The mass M of the group is distributed over the ball giving the average density

$$\langle \rho_M \rangle = \frac{M}{V_\Lambda} = 2\rho_\Lambda$$

leading to the ratio (6.3) equal exactly to unity, independent of the group mass,

$$X_{LG} = 1 . \tag{6.4}$$

This is just another way of stating that the amounts of attractive and repulsive mass evenly distributed inside the no-gravity sphere are equal, see Eq. (2.8).

If the effective radius of a local group ball is somewhat different from r_Λ, and/or its shape deviates from the spherical one, then X_{LG} would be different, but still in the ballpark of $X_{LG} \sim 1$.

The universal result (6.4) should work for many clusters, because their shape is often close to a spherical one. Particularly, this prediction was confirmed in [12] for the *Virgo*, *Fornax*, and *Coma* clusters of galaxies whose mass is 7–10 times larger than the mass of the Local Group. This is also a significant confirmation of the first part of our general conjecture.

6.3.2 Local Pancake (2D)

The Local Pancake is a supercluster of about 20 Mpc across; we call it the Zeldovich Local Pancake (ZLP) honoring Ya. B. Zeldovich who theoretically established the existence of pancakes in his famous paper [41]. ZLP is seen as an expansion flow of giant galaxies, which was studied with the HST by Karachentsev et al. (see [15] and the references therein). The flow involves 15 most luminous nearby galaxies together with about 300 their fainter companions.

ZLP occupies a flattened (disc-like) volume located near the Supergalactic Plane. The flow is expanding: the giants are moving away from the barycenter of the group with the radial velocities from 100 up to 1000 km/s. The total matter mass of the ZLP is $M \simeq 8 \times 10^{13} M_\odot$. The density ratio is definitely and significantly larger than unity,

$$X_{ZLP} \geq 10 . \tag{6.5}$$

This means that DE antigravity dominates the dynamics of the ZLP, its flow of galaxies is expanding with acceleration. The change of the expansion rate can be characterized by the dimensionless deceleration parameter $q(r) = -\ddot{r}r/v^2$ borrowed from cosmology. The parameter turns out negative for each of the ZLP individual galaxies at the present epoch of observations, demonstrating that they are accelerating. Its mean value for the flow as a whole is $\langle q(r) \rangle \simeq -0.9$. This is an additional method to quantify the DE domination; similar values of q may be expected in other typical expanding superclusters.

6.3.3 Filament Feeding a Massive Cluster (1D)

Weak lensing detection of a large-scale filament funneling matter onto a core of the massive galaxy cluster MACSJ0717.5+3745 was reported recently in [42]. The proper length of the filament is 18 Mpc, and the mean matter density is determined to be 2×10^{-27} g/cm^3. Accordingly, the ratio X for the filament is about 0.01 $\ll$ 1, meaning that DE effects on the filament dynamics are negligibly small, and confirming the third part of our conjecture. There are all grounds to expect that most large filaments are also strongly dominated by matter gravity.

So, the basic building blocks of the Cosmic Web, and the Web as a whole are embedded in the universal dark energy background. The background produces antigravity which is often strong inside clusters ($1 \lesssim X$) and superclusters (pancakes) ($X \gg 1$) of galaxies; such strong antigravity effect is reliably observed inside typical systems of these kind. Contrary to that, antigravity is very weak ($X \ll 1$) in large filaments, where gravity produced by matter (dark matter and baryons) prevails. The gravity-antigravity interplay is one of the most impressive phenomena predicted theoretically by General Relativity with the cosmological constant; it reveals itself in various modern observations of the Cosmic Web.

Chapter 7
Concluding Remarks

Abstract Final comments on fascinating properties of dark energy and some open questions about its nature.

We discussed a number of theoretical and observational issues related to a wonderful and still mysterious thing, dark energy, sometimes also called heavy vacuum, whose history is revealed in the book [43]. It was introduced by Einstein and got its physical interpretation from Gliner; for this reason, in [4] we suggested yet another name for it, the Einstein–Gliner vacuum, or shortly EG-vacuum. In physics there exists a plethora of vacua: 'technical' vacuum , true vacuum, false vacuum, quantum vacuum, and so on; however, EG-vacuum is truly unique against this rich background.

EG-vacuum is a special medium, unknown in physics before. It is not visible, does not emit, absorb, or scatter light. But it is not just emptiness, because EG-vacuum has a non-zero energy. This energy fills the whole universe perfectly uniformly: the vacuum energy density is the same everywhere and does not change with the time.

EG-vacuum also has pressure, and the relation between it and the energy density, which, for any medium, is called the equation of state, is: the pressure is the energy density with a negative sign, as stated in Sect. 2.1. The vacuum energy density is positive, hence its pressure is negative. Negative pressure is not so unusual: for instance, the pressure is negative inside a stretched rubber tourniquet, or in a steel block stretched in all directions. However, it is only in EG-vacuum, and not in any other medium, that the pressure is equal in magnitude to the energy density.

Two most important properties of EG-vacuum follow from its equation of state. First, it enjoys the main mechanical property of 'vacuum', namely, that motion and rest relative to it cannot be discerned (as in the trivial case of emptiness). Alternatively, one can describe this as follows. Let there be two bodies (or two frames) that move against each other in an arbitrary way. However, both bodies (or frames) are always at rest relative to vacuum. So EG-vacuum co-moves with any body. Clearly, it cannot serve as a frame, for this reason, and its energy density is the same in all frames, whatever their relative motion is. In yet other words, EG-vacuum is always and everywhere the same in all its manifestations.

Second, EG-vacuum creates not the attraction but repulsion, the overall repulsion of the universe. This is because in General Relativity the ability of a uniform medium

A. Silbergleit and A. Chernin, *Kepler Problem in the Presence of Dark Energy,
and the Cosmic Local Flow*, SpringerBriefs in Physics,
https://doi.org/10.1007/978-3-030-36752-7_7

to create gravity depends not only on its density, as in Newton's theory, but also on its pressure. In cosmology, the 'effective gravitating energy density' is a combination of these two quantities: it is the sum of the energy density and three values of pressure. For vacuum, with its special equation of state, this sum is negative, equal to negative twice density, as shown by Eq. (2.2). That is, unlike all other substances in nature, EG-vacuum creates anti-gravity instead of gravity.

There exists a large variety of questions regarding EG-vacuum, or dark energy. The most important of them is about its internal structure. Neither Gliner's ideas, nor astronomical observations tell us anything about the content of the vacuum at the microscopic level. What kind of substance is vacuum? What is it 'made of'? The question remains completely unanswered. Many people think that this is a key problem of fundamental physics in the 21st century.

We touched upon the other very important question in the previous chapter: does dark energy interact with matter, or it does not? In the latter case it is exactly the universal EG-vacuum whose density does not vary in space and time. Whether, instead, the former case actually takes place, i.e., DE and matter interact with each other, can hopefully be established through astronomical observations. The detection of some macroscopic effects of DE-matter interaction might eventually help with clarifying the fundamental problem of the DE microscopic structure, as well. James Bjorken has repeatedly suggested trying to come up with a map of the dark energy distribution in the universe. From this point of view, the work on the local cosmic flows discussed in Chap. 6 can be considered as the first small step towards the creation of such map. So far, the results point to the universality of dark energy, providing the local values of its density close to the one determined at cosmological scales.

Acknowledgements A. S. thanks the participants of Gravity Probe B Theory Group for discussing his talk on Kepler problem with dark energy. A. Ch. thanks I. D. Karachentsev, N. V. Emelyanov, and G. S. Bisnovatyi-Kogan for their collaboration and fruitful discussions. We are grateful to Kseniya Makarova for her help in preparing figures.

Appendix A
Radial Solution in Elliptic Integrals for $E < U_{max} < 0$

According to Sect. 3.5.1, we need to reduce the integral

$$Ht = \sqrt{b} \int_u^1 \frac{du}{u\sqrt{(u+\xi)(u^2 - Au + \xi A)}} \equiv \sqrt{b}I, \quad A = \xi - a > 0,$$

to a combination of standard elliptic integrals; we do it in three steps.

First, we use the shift of the integration variable,

$$z(u) = u - A/2, \tag{A.1}$$

to get rid of the linear term is the quadratic polynomial, resulting in

$$I = \int_{z(u)}^{z(1)} \frac{dz}{(z + A/2)\sqrt{(z + A/2 + \xi)(z^2 + q^2)}}. \tag{A.2}$$

Here we denoted, for brevity,

$$q^2 = A^2/4 - A^2/2 + \xi A = (A/4)(4\xi - A) = (\xi - a)(3\xi + a)/4 > 0. \tag{A.3}$$

Next we exploit the Euler substitution,

$$\sqrt{z^2 + q^2} = s + z, \quad s(u) = \sqrt{(u - A/2)^2 + q^2} - (u - A/2), \tag{A.4}$$

with the formula for $s(u)$ implied by the expression (A.1) for $z(u)$. This is done to obtain a quadratic polynomial with real roots. After the necessary algebra, our integral (A.2) converts to

© The Author(s), under exclusive license to Springer Nature Switzerland AG 2019
A. Silbergleit and A. Chernin, *Kepler Problem in the Presence of Dark Energy, and the Cosmic Local Flow*, SpringerBriefs in Physics,
https://doi.org/10.1007/978-3-030-36752-7

$$I = 2^{3/2} \int_{s(1)}^{s(u)} \frac{s\,ds}{(-s^2 + As + q^2)\sqrt{s\left[-s^2 + (A + 2\xi)s + q^2\right]}} \equiv$$

$$2^{3/2} \int_{s(1)}^{s(u)} \frac{s\,ds}{y_2(s)\sqrt{sy_1(s)}} , \tag{A.5}$$

with two new quadratic trinomials $y_1(s)$ and $y_2(s)$. They both have real roots: in case of $y_1(s)$, which is crucial for the latter derivation, they are

$$s_\pm = \frac{1}{2}\left[(3\xi - a) \pm 2\sqrt{\xi(3\xi - 2a)}\right], \quad s_- < 0 < s_+, \quad 0 < \frac{s_+}{s_+ - s_-} < 1, \quad (A.6)$$

so

$$y_1(s) = (s_+ - s)(s - s_-) > 0 \tag{A.7}$$

for s between s_- and s_+, as in the integral (A.5). Similarly, the roots of $y_2(s)$ are

$$c_\pm = \frac{1}{2}\left[-(\xi - a) \pm 2\sqrt{\xi(\xi - a)}\right], \quad c_- < 0 < c_+, \quad c_+ < s_+ , \tag{A.8}$$

and

$$y_2(s) = (c_+ - s)(s - c_-) .$$

Therefore we can write the integral (A.5) as

$$I = 2^{3/2} \int_{s(1)}^{s(u)} \frac{s\,ds}{(c_+ - s)(s - c_-)\sqrt{s(s_+ - s)(s - s_-)}} ,$$

or, after breaking the rational part down to simple fractions,

$$I = 2^{3/2} \times$$

$$\left[\int_{s(u)}^{s(1)} \frac{D_+\,ds}{(s - c_+)\sqrt{s(s_+ - s)(s - s_-)}} - \int_{s(u)}^{s(1)} \frac{D_-\,ds}{(s - c_-)\sqrt{s(s_+ - s)(s - s_-)}}\right] ,$$

$$\tag{A.9}$$

where $D_\pm = c_\pm/(c_+ - c_-)$.

The third and last step is to transform both integrands to the Legendre form,

$$\frac{1}{(1 + \nu x^2)\sqrt{(1 - x^2)(1 - k^2 x^2)}} .$$

This is done by the substitution [see (A.1), (A.4)]

$$\sqrt{s_+ - s} = \sqrt{s_+}\, x \,, \tag{A.10}$$

$$x(u) = \sqrt{1 - (1/s_+)\left[\sqrt{(u - A/2)^2 + (3\xi + a)(A/4)} - (u - A/2)\right]} \,,$$

that leads precisely to the solution (3.14).

A good way to check our calculations is to demonstrate the validity of the Hubble law (3.7) in the limit $u \to +0$ ($r \to \infty$). To identify the leading singular term in the expression (3.14) we note that, by (A.10) and the expression (3.14) for ν_+,

$$x^2(0) = (s_+ - c_+)/s_+ = -1/\nu_+ \,.$$

Hence the leading term comes from the first elliptic integral in (3.14). After cumbersome but simple manipulations we find it to be

$$Ht = -K \int_{x(u)}^{x(1)} \frac{dx}{x(0) - x} = K \int_u^1 \frac{du}{u} = -K \ln u = K \ln \frac{r}{r_0} \,,$$

since to l.o. $x(0) - x = -x'(0)u$, $dx = x'(0)du$ in the limit $u \to 0$. Here

$$K = c_+ \sqrt{\frac{2\xi}{s_+(s_+ - s_-)(s_+ - c_-)[1 - x^2(0)][1 - k^2 x^2(0)]}} \,,$$

and we only need to prove $K = 1$ to establish the agreement with the leading term of the Hubble law (3.7).

Using the above expression for $x^2(0)$ and the formula (A.7), we find

$$K = \sqrt{\frac{2\xi c_+}{(c_+ - s_-)(s_+ - c_+)}} = \sqrt{\frac{2\xi c_+}{y_1(c_+)}} \,.$$

By the definition (A.5) of the two quadratic polynomials $y_1(s)$ and $y_2(s)$,

$$y_1(s) = -s^2 + (A + 2\xi)s + q^2 = (-s^2 + As + q^2) + 2\xi s = y_2(s) + 2\xi s \,.$$

However, c_+ is the root of $y_2(s)$, so

$$y_1(c_+) = y_2(c_+) + 2\xi c_+ = 2\xi c_+ \,,$$

and $K = 1$, as required.

Appendix B
Asymptotic Expressions for the Roots of $Q(u)$ and Critical Parameters a', a'', or E', E''

We derive asymptotic approximations for the four real roots of $Q(u)$ that exist when $a' \leq a \leq a'' < 0$ and $0 < b < b_*$. The equation

$$Q(u) = -lu^4 + u^3 + au^2 + b = 0$$

under the change of the unknown and parameter

$$u = -w/l, \quad a = -\alpha/l, \quad 0 < \alpha' \leq \alpha \leq \alpha'' < 9/32 \qquad (B.1)$$

becomes

$$-\frac{1}{l^3}(w^4 + w^3 + \alpha w^2) + b = 0 \,,$$

or

$$w^2(w^2 + w + \alpha) = \beta \,, \qquad (B.2)$$

where $\beta = l^3 b$ is a small positive parameter defined in Sect. 4.3.2, $0 < \beta < (3/16)^3$. Therefore it is natural to solve this equation by perturbations in β.

For $\beta = 0$ the unperturbed equation (B.2) is

$$w^2(w^2 + w + \alpha) = 0 \,.$$

It has a double root $w = 0$, which splits under perturbation in two, one negative and one positive, corresponding to $-\xi$ and η; the expansion goes in half-integer powers of β, and the result is (see notations (B.1)):

$$-\xi = \frac{1}{l\alpha^{1/2}}\left(\beta^{1/2} + \frac{1}{\alpha^{3/2}}\beta^{3/2} + \cdots\right); \quad \eta = \frac{1}{l\alpha^{1/2}}\left(\beta^{1/2} - \frac{1}{\alpha^{3/2}}\beta^{3/2} + \cdots\right); \quad (B.3)$$

it is not difficult to obtain higher order terms as well, but they are hardly needed in practice, since $\beta \lesssim 10^{-3}$.

© The Author(s), under exclusive license to Springer Nature Switzerland AG 2019
A. Silbergleit and A. Chernin, *Kepler Problem in the Presence of Dark Energy, and the Cosmic Local Flow*, SpringerBriefs in Physics,
https://doi.org/10.1007/978-3-030-36752-7

The other two roots of the unperturbed equation are:

$$w_{1,2} = -\frac{1}{2}\left(1 \mp \sqrt{1 - 4\alpha}\right),$$

so they are real for $\alpha \geq 1/4$. Since, as seen from Fig. 4.6a, b, the real roots appear when growing a reaches a' (decreasing α reaches α'), this immediately provides the lowest order value for the latter to be one quarter,

$$\alpha' = 1/4 ; \qquad (B.4)$$

the correction to this is found below.

If $\alpha = 1/4$, then the unperturbed equation has a double root $w = -1/2$; similar to the double root $w = 0$, it splits in two under perturbation with the expansion going in half-integer powers of β; in terms of the original unknown u, those expansions are:

$$u_1 = \frac{1}{2l}\left(1 - 4\beta^{1/2} + \cdots\right), \qquad u_2 = \frac{1}{2l}\left(1 + 4\beta^{1/2} + \cdots\right), \qquad \alpha = \frac{1}{4}. \quad (B.5)$$

If $\alpha > 1/4$, then each of the roots is simple, the expansion goes in integer powers:

$$u_1 = \frac{1}{2l}\left[\left(1 - \sqrt{1 - 4\alpha}\right) - \frac{2}{\sqrt{1 - 4\alpha}\left(1 - 2\alpha - \sqrt{1 - 4\alpha}\right)}\beta + \cdots\right], \qquad (B.6)$$

$$u_2 = \frac{1}{2l}\left[\left(1 + \sqrt{1 - 4\alpha}\right) + \frac{2}{\sqrt{1 - 4\alpha}\left(1 - 2\alpha + \sqrt{1 - 4\alpha}\right)}\beta + \cdots\right], \qquad \alpha > \frac{1}{4}.$$

Note that all the asymptotic expansions (B.3), (B.5), and (B.6) can be obtained by the general method for algebraic perturbations, the Newton diagram [44, 45].

We now turn to the asymptotic approximations for a' and a'', or α' and α'. To get them from their definitions (4.15), (4.16) one needs to simultaneously solve for u and a two equations,

$$Q(u) = -lu^4 + u^3 + au^2 + b = 0 \quad \text{and} \quad Q'(u) = -4lu^3 + 3u^2 + 2au = 0 ;$$

a solution with smaller negative a provides a', and the other with larger negative a delivers α''. Turning to the parameters w and α according to (B.1) gives ($w \neq 0$):

$$w^4 + w^3 + \alpha w^2 = \beta, \qquad 4w^2 + 3w + 2\alpha = 0 .$$

Expressing α from the second equation,

$$\alpha = -\left(\frac{3}{2}w + 2w^2\right), \qquad (B.7)$$

and introducing it to the first one we obtain the single resolving equation for w:

$$w^4 + \frac{1}{2}w^3 = -\beta .$$

The corresponding unperturbed ($\beta = 0$) equation has a single root $w = -1/2$ and a triple root $w = 0$. The first one gives rise to a regular expansion in integer powers of β, the second splits in three roots as $\beta^{1/3}$, with just one of them being real:

$$w_1 = -\frac{1}{2} + 8\beta + \cdots , \qquad w_2 = -2^{1/3}\beta^{1/3} - \frac{2^{5/3}}{3}\beta^{2/3} + \cdots .$$

This being substituted in the formula (B.7) allows for

$$\alpha' = \frac{1}{4} + 4\beta + \cdots , \qquad a' = -\frac{1}{l}\left(\frac{1}{4} + 4\beta + \cdots\right) ; \qquad \text{(B.8)}$$

$$\alpha'' = \frac{3}{2}(2\beta)^{1/3} - (2\beta)^{2/3} + \cdots , \qquad a'' = -\frac{1}{l}\left[\frac{3}{2}(2\beta)^{1/3} - (2\beta)^{2/3} + \cdots\right] ,$$

which is the desired result. In terms of energy it looks as

$$E' = -\frac{GM}{lr_0}\left(\frac{1}{4} + 4\beta + \cdots\right) ; \qquad E'' = -\frac{GM}{lr_0}\left[\frac{3}{2}(2\beta)^{1/3} - (2\beta)^{2/3} + \cdots\right] ,$$

$$\text{(B.9)}$$

according to the formulas (4.18).

References

1. E.B. Gliner, Sov. Phys. JETP **22**, 378 (1965)
2. E.B. Gliner, Sov. Phys. Dokl. **15**, 559 (1970)
3. E.B. Gliner, I.G. Dymnikova, Sov. Astron. Lett. **1**, 93 (1975)
4. A.S. Silbergleit, A.D. Chernin, *Interacting Dark Energy and the Expansion of the Universe.* Springer Briefs in Physics (Springer, Berlin, 2017)
5. A. Guth, Phys. Rev. D **23**, 347 (1981)
6. E.B. Gliner, Phys. Usp. **45**, 213 (2002)
7. A.D. Chernin, P. Teerikorpi, Yu.V. Baryshev, Why is the Hubble flow so quiet? Talk at COSPAR-2000, Warsaw, 2000. Published in: Adv. Space Res. **31**, 459 (2003)
8. A.D. Chernin, Phys. Usp. **44**, 1099 (2001)
9. Yu. Baryshev, A. Chernin, P. Teerikorpi, Astron. Astrophys. **378**, 729 (2001)
10. I.D. Karachentsev, A.D. Chernin, P. Teerikorpi, Astrofizika **46**, 491 (2003)
11. I.D. Karachentsev, ApJ **129**, 178 (2005)
12. G.S. Bisnovatyi-Kogan, A.D. Chernin, Astrophys. Space Sci. **338**, 337 (2012)
13. A.D. Chernin, Phys. Usp. **56**, 704 (2013)
14. N.V. Emelyanov, M.Yu. Kovalyov, MNRAS **429**, 3477 (2013)
15. A.D. Chernin, N.V. Emelyanov, I.D. Karachentsev, MNRAS **449**, 2069 (2015)
16. N.V. Emelyanov, M.Yu. Kovalyov, A.D. Chernin, Astron. Rep. **59**, 510 (2015)
17. A.D. Chernin, Astron. Rep. **59**, 474 (2015)
18. N.V. Emelyanov, M.Yu. Kovalyov, A.D. Chernin, Astron. Rep. **60**, 397 (2016)
19. M. McLeod, O. Lahav (2019), arXiv:1903.10849
20. G.S. Bisnovatyi-Kogan, M. Merafina (2019), arXiv:1906.05861
21. F.I. Cooperstock, V. Faraoni, D.N. Vollick, ApJ **503**, 61 (1998)
22. A. Dominguez, J. Gaite, Europhys. Lett. **55**, 458 (2001)
23. S. Nesseris, L. Perivolaropoulos, Phys. Rev. D **70**, 123529 (2004)
24. G.S. Adkins, J. McDonnell, R.N. Fell, Phys. Rev. D **75**, 064011 (2007)
25. M. Sereno, P. Jetzer, Phys. Rev. D **75**, 064031 (2007)
26. V. Faraoni, A. Jacques, Phys. Rev. D **76**, 063510 (2007)
27. B. Mashhoon, N. Mobed, D. Singh, Class. Quantum Gravity **24**, 5031 (2007)
28. M. Carrera, D. Giulini, Rev. Mod. Phys. **82**, 169 (2010)
29. A.D. Chernin et al., Astron. Rep. **54**, 185 (2010)
30. L.D. Landau, E.M. Lifshitz, *Mechanics* (Pergamon Press, Oxford, 1959)
31. H.B. Dwight, *Tables of Integrals* (Macmillan, New York, 1961)

© The Author(s), under exclusive license to Springer Nature Switzerland AG 2019
A. Silbergleit and A. Chernin, *Kepler Problem in the Presence of Dark Energy, and the Cosmic Local Flow*, SpringerBriefs in Physics,
https://doi.org/10.1007/978-3-030-36752-7

32. H. Bateman, A. Erdélyi, *Higher Transcendental Functions*, vol. 3 (McGraw-Hill Book Co., New York, 1955)
33. A.A. Friedmann, Zeitschrift für Physik **21**, 326 (1922)
34. A.S. Silbergleit, Astron. Astrophys. Trans. **21**, 171 (2002)
35. J. Einasto, *Dark Matter and Cosmic Web Story* (World Scientific, Berlin, 2014)
36. W.M. Irvine, Local irregularities in a universe satisfying the cosmological principle. Ph.D. thesis, Harvard University, Cambridge (1961)
37. D. Layzer, ApJ **138**, 174 (1963)
38. N.A. Dmitriev, Ya.B. Zel'dovich, JETP **18**, 793 (1963)
39. A.D. Chernin et al., Astron. Rep. **54**, 185 (2010)
40. A.S. Silbergleit, A.D. Chernin, in *Proceedings of the 4th International Conference on Topical Problems of Mechanics of Continua, 21–25 Sept 2015, Tsahkadzor, Armenia* (2015), pp. 496–498. Full text of the talk is found at: https://www.researchgate.net/profile/AlexanderSilbergleit/research
41. Ya.B. Zeldovich, Astron. Astrophys. **5**, 84 (1970)
42. E. Jauzac et al., MNRAS **426**, 3360 (2014)
43. H. Kragh, J. Overduin, *The Weight of the Vacuum: A Scientific History of Dark Energy*. Springer Briefs in Physics (2014)
44. http://encyclopediaofmath.org/index.php/Newtondiagram
45. M.M. Vainberg, V.A. Trenogin, *Theory of Branching of Solutions of Non-linear Equations* (Noordhoff, Leyden, 1974)

Printed in the United States
By Bookmasters